Introduction to Integrated
Geo-information Management

Introduction to Integrated Geo-information Management

Seppe Cassettari Phd

Course Director for GIS studies
Kingston University
Kingston-upon-Thames, UK

CHAPMAN & HALL
London · Glasgow · New York · Tokyo · Melbourne · Madras

Published by Chapman & Hall, 2-6 Boundary Row, London SE1 8HN

Chapman & Hall, 2-6 Boundary Row, London SE1 8HN, UK

Blackie Academic & Professional, Wester Cleddens Road, Bishopbriggs, Glasgow G64 2NZ, UK

Chapman & Hall Inc., 29 West 35th Street, New York NY10001, USA

Chapman & Hall Japan, Thomson Publishing Japan, Hirakawacho Nemoto Building, 6F, 1-7-11 Hirakawa-cho, Chiyoda-ku, Tokyo 102, Japan

Chapman & Hall Australia, Thomas Nelson Australia, 102 Dodds Street, South Melbourne, Victoria 3205, Australia

Chapman & Hall India, R. Seshadri, 32 Second Main Road, CIT East, Madras 600 035, India

First edition 1993

© 1993 Seppe Cassettari

Printed in Great Britain by T.J. Press (Padstow) Ltd, Padstow, Cornwall

ISBN   0 412 48900 7

A catalogue record for this book is available from the British Library

Library of Congress Cataloging-in-Publication data available

∞ Printed on permanent acid-free text paper, manufactured in accordance with ANSI/NISO Z39.48-1992 and ANSI/NISO Z39.48-1984 (Permanence of Paper).

Contents

Preface

This book was inspired by the revolution in geographical information systems during the late 1970s and 1980s which introduced to many the concept of computer-based information systems for spatially referenced data. The map, the aerial photograph and the satellite image were wedded to a database of textual information through the rapidly developing technology of powerful graphics workstations. This brought the skills of the geographer to a wide range of disciplines and specialists.

But this book is not about the basic concepts of geographical information systems themselves. It is not about hardware or software per se, nor the integral concepts of geo-referenced data handling built into such systems; these are to be found in a growing number of introductory texts on the subject. Instead the focus of this book is on the much wider issues of geo-information management.

While an understanding of the systems, their capabilities and limitations is necessary, of greater importance to the long term application of geographical understanding to problem solving is the wider context of information handling. Spatial data are becoming increasingly important in understanding the issues that confront the world.

Chapter 1 is a discussion of the general issues which relate to management and information systems. It concludes with review of spatial decision support systems which are of increasing importance to the GIS community.

The second chapter is concerned with data and the issues that relate to quality and accuracy. It includes a comment on the need for metadata within spatial information systems. Chapter 3 looks at the database issues for the storage and management of spatial data while Chapters 4 and 5 address the key areas of system integration. The first of these chapters concentrates on large corporate systems and the latter addresses some of the specific points that relate to low cost solutions.

In the sixth chapter some of the legal and copyright issues are discussed, highlighting the need for greater awareness of data ownership and greater accountability in terms of the data generated from a GIS. In Chapter 7 standards are reviewed, including a summary of the current data standards.

Chapter 8 considers the role of analysis in geo-information management, while Chapter 9 looks at user interfaces and their importance in effective information dissemination. In Chapter 10 the problems of visualisation are considered with the emphasis on the need for the GIS community to learn from the work of the cartographic industry.

The last two chapters look at new and potentially important technologies which will play an increasing role in the development of integrated solutions. These are image-based spatial information systems, which use aerial photography and satellite imagery rather than maps and multimedia GIS. This last includes a review of hypertext and hypermap concepts.

The book is intended to complement those texts which provide an introduction to the basic concepts of GIS technology and to bridge the gap between GIS and the broader field of information management.

Seppe Cassettari

Acknowledgements

That this book was completed and on time is a tribute to the GIS staff at the School of Geography, Kingston University, who both supported the work and contributed valuable ideas and comments. My thanks to all of them, but especially Tim Lindsey, Ed Parsons and Alun Jones for their very useful comments and suggestions and Debbie Millard for the excellent graphics.

Special thanks also go to Steve Dunnico for the cover illustration and to my wife Sylvia for her support.

List of figures

List of tables

1

Information management

1.1 Introduction

The twentieth century has been marked by an explosion in the availability of information. Huge amounts of data are collected to satisfy our demand for knowledge and understanding of the world in which we live, to exploit the resources necessary to provide the infrastructure of human life and to manage our societies. The revolution in information technology has done much to influence these activities. The development of digital computer systems and sophisticated methods of communication have had a profound effect on the way individuals and organisations work, how we use our leisure time and interact with others from a local to a global scale.

The pace of change is fast and increasing. New technologies are becoming available all the time as part of the overall economic objective of society for continuing growth and wealth creation. The developments in audio reproduction from long playing records to compact discs provide one example, the rapid developments in television and video provide another. The computer has changed from being a room full of electronic hardware to a sophisticated machine sitting on a desk. During that time the power of the computer has increased a hundred fold and the costs have declined proportionally. Such changes are to meet the increasing demands for appropriate and viable tools to do increasingly complex tasks and for those involved in the development of information technology to continually find new products to sell and new markets to sell to.

The growth in the use of information handling technology is usually driven by technical developments, rarely has it been determined by social needs. As a consequence once a new technical development becomes available only then are the application areas fully addressed. In some cases a new technology has spawned a new social activity or created a new work environment. The way many businesses conduct themselves has changed dramatically with the introduction of the fax machine, and the developments in the personal computer are having a profound effect on leisure activities in the home, with young people turning away from traditional board games to interative computer games. The most successful innovations in information technology cause some form of cultural or attitude change by individuals or organisations which lead to the embracement of that technology and the perceived wisdom 'how did we ever manage without it?'

Often the last aspect to be addressed by the introduction of new information technology is the social or management consequences. That society in general or a particular organisation will have to adjust as a result of such developments is apparent from the necessary cultural or attitude changes the particular technology brings. However, such change is often slow and the knock-on effects can be unpredictably complex. Issues such as availability of information, personal freedoms, confidentiality and security all have far reaching implications as technology makes the collection and storage of data more efficient with potentially wider and more immediate access to it.

The computer industry in particular is characterised by rapidly changing capabilities. The applications for the technology often lag some way behind new developments and a last consideration is usually the management and organisational changes affected by these technologies. This problem has an impact at a broad social level and all the way down to the individual level. In particular it is often hard to identify the impact of a technology on an organisation or on the work routine or social life of an individual. There are many organisations where technical innovations have not been successful for organisational and management reasons even though the technology could have been of direct and lasting value. This problem is one of the major concerns of those involved in spatial information handling.

1.2 Spatial data systems

The development of a conceptual framework and appropriate technology for spatial data handling started in the 1960s but advanced rapidly in the 1980s and early 1990s with the increase in computer capabilities. Now the geographical information systems (GIS) industry brings together a wide range of disciplines but is founded on the basic geographical concept that many of the processes of society and the global environment are determined by spatial interaction.

GIS encompasses information technology, information management, business and legal issues and subject specific concepts from a wide range of disciplines, but it is implicit in the idea of GIS that it is technology used to support decision making for solving problems that in part at least have a spatial component (Maguire, 1991).

As a tool GIS includes data acquisition and capture into computer format, the appropriate management of that data within a computer environment, the analysis of data in various ways and the output of the results. All these elements are essential in GIS systems but the relative emphasis of each aspect will depend on the purpose for which the system was originally designed.

There are a great many software packages available which are grouped under the generic heading of GIS but which do not contain all these basic elements, in particular they lack analytical capabilities. These are more properly termed digital mapping or computer-aided design (CAD) packages.

It is not always possible to make a clear distinction between the various categories, especially as system functionality increases with product development. The large number of solutions on the market are best viewed in terms of points along a continuum, with simple geo-referenced drawing capabilities at one extreme and highly sophisticated management and analytical functions at the other. The available systems for low-cost platforms are discussed in section 5.2 with a summary produced as Appendix B. A fuller review of all the main GIS systems and their capabilities may be found in Enyon (1992).

The geographical information system is a relatively 'young' addition to the information technology portfolio, and as the name implies, is still driven by the development of computer hardware and software. Only in recent years have large numbers of applications been developed which integrate the spatial dimension with other information handling technologies, and it has yet to achieve the organisational and cultural changes required to ensure it becomes an integrated part of the wider information society.

This book addresses the role of GIS in a broad information handling context and as part of an integrated decision-making structure. It will look more particularly at the management and organisational issues as they are affected by developing conceptual and technical innovation. It will not address the technical details of data handling such as data structures and analytical techniques, assuming the reader has a basic level of knowledge and understanding. Those who feel the need of this background should look at one of the introductory texts on GIS, for example Burroughs (1986), Aranoff (1989) or Laurini and Thompson (1992).

1.3 Decision making

The process by which individuals make decisions is a complex one. Different people when faced with the same problem will adopt different approaches to its solution, will assign varying priorities to the individual elements and their inter-relationships, and will select and use information in different ways. We all make decisions constantly and measure the results of those decisions either consciously or unconsciously, depending on the required and achieved results. The decision-making process is controlled by a number of factors:

- The regulations which define the constraints within which the decision is made
- The various and changing circumstances pertaining at the time
- An individual's experience of similar decision-making situations
- Knowledge about the desired outcomes

For example, consider the case of someone wishing to cross a road. There may be legal constraints that stop the individual from crossing at a particular point but outside of these they may choose the best crossing place. A young child may wish to get to the other side of the road but lack the experience to identify a suitable crossing point. Adults on the other hand will use their experience to take into account speed and amount of traffic, obstructions and places where they can not see clearly and as a result choose to cross using a pedestrian crossing located for that purpose. They may on the other hand decide to take a greater risk and dodge between cars.

Crossing successfully to the other side is the measure by which the decision to cross at a particular place and time is deemed to have been the correct one. There are a number of measures by which an individual may deem the decision to be the wrong one, such as an accident or a heart-stopping near miss. More often than not there will be no conscious evaluation of the original decision. Also experience in one environment is not always easily translated to another, for example the problems faced by an individual trying to cross the road in a country with vehicles driving on the left when they are used to vehicles driving on the right.

Experience, the control imposed by laws and regulations and pertaining circumstances are all important to the decision-making processes in a work environment. The relative importance of each depends on the nature of the decision and it is necessary to establish decision-making frameworks within which organisations operate. At one level individuals have a responsibility to manage their own work environment, while at the highest level decisions are made about the nature and objectives of that organisation. All activities within an organisation can be seen as 'a series of decisions and even at the lowest levels of skill and experience decisions are being made constantly' (Anderson, 1986). It is the organisational structure and the rules by which it operates that determines the scope and impact of any one decision.

An important part of structuring an organisation is determining where within the structure decisions are made. Within a typical company there are clearly decisions which are made on a day to day basis by individuals or those who head small sections, while at the other end of the scale there are decisions which are the province of the senior management or of the boardroom. It is necessary to delegate decision making to the level at which the appropriate information is available and where the resulting action can be taken. These may not be one and the same, in which case the organisation has to question its structure and the flow of information within it (Figure 1.1).

Those responsible for an organisation have to balance the need to maintain control while at the same time achieving flexibility and efficiency. The delegation of decision-making powers to an inappropriate level can lead to a loss of control where managers do not know what is occurring within the organisation. On the other hand retaining decision making at too high a level can place great strain on particular individuals or information nodes within the organisation and lead to a low level of staff involvement in the organisation.

The distinction may be made between strategic and tactical decisions. Where decisions have a significant effect on attaining the identified goals of the organisation, in particular aspects of resourcing, these may be regarded as strategic and are taken at a senior level. These policy decisions determine the range of actions that progressively filter down through an organisation to a level where the appropriate decisions are taken in response to the defined policy and overall goals.

Computer tools exist to support decision making processes. These include management information systems (MIS), decision support systems (DSS) and executive support systems (ESS). These are discussed further in the context of spatial information in section 1.9.

However, as Anderson (1986) rightly points out, decision-making is not simply a sequence of planning, policy-making and various stages of operating system leading to action. There are quantitative decisions such as what to buy, what price to pay and how many, and there are qualitative decisions such as who should do the job. As a result even the simplest of tasks involve complex decisions and rarely go completely according to plan. Those with the decision-making responsibility are constantly having to review their plans and adjust the decisions to the prevailing conditions. There are of course a wide range of eventualities that can be largely overlooked or unexpected, with the result that managers are often making 'instant' decisions in response to changes in circumstances. In the worst situation decisions are made with little or no planning or forethought and result in 'fire fighting' - no sooner is one problem solved than two or more raise their heads and call for instant

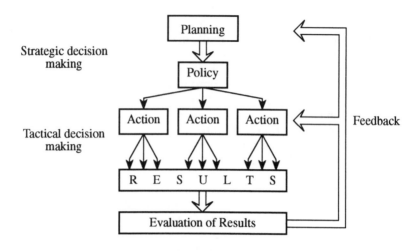

Figure 1.1 Decision-making structure.

responses.

In the more successful decision-making structures an element of review concerning past decisions is integrated into the whole process. This provides a feedback mechanism by which the results from previous decisions can be assessed. This is then fed into the planning process and should lead to improved decision making in the future. Feedback mechanisms for decision-making processes are often difficult to establish and may well be the first casualty of increased pressure on the organisation to achieve particular targets. In this way poor decisions can be perpetuated and small problems that derive from them in the first instance can develop into major problems at a later stage.

The most important factor in the whole decision-making process is the human component. This is the most unpredictable and most changeable. The problems of managing people and how to match the human resource with the goals set for the organisation present the greatest challenge to the decision-making structure. The impact of training not only in the task but also for those involved in human resource management is particularly important. Developing the skills necessary for successful decision making and how to evaluate results is a goal many organisations set for individuals in order to improve their overall performance.

1.4 Organisations and their objectives

The purpose of decision making is to meet the objectives that an organisation or business sets itself. Every organisation sets overall objectives and internally different component parts of the organisation will also have their own objectives which may, in part at least, differ substantially from those of the organisation as a whole. It is worth pointing out that an 'organisation' can encompass central and local governments or parts of government, commercial businesses, social or cultural structures such as schools, universities or hospitals, and non-governmental pressure groups and charities. Thus the term organisation needs to be viewed in its widest possible context.

Objectives may be categorised into strategic objectives which determine the overall direction of the organisation and secondly the financial and resource targets necessary to achieve the strategic goals. Anderson (1986) identified five ways in which strategic level objectives may be defined:

- Profit - A measure by which the organisation or business has achieved a greater return on the work it has carried out compared to the resources committed. There are many ways in which profit may be measured, not just in financial terms, and there are various timescales which may also have to be taken into account.
- Survival - The primary concern of many organisations is to continue operating and the primary decisions will be to ensure financial stability. Any profit will be regarded as a bonus.
- Expansion - Expansion, while linked to survival and profit, is a route often sought as a means to survival by a business that is struggling. Growth is a natural inclination of many organisations and often failure to achieve growth is viewed as failure.
- Equilibrium - The desire to retain the current status quo and maintain equilibrium is an alternative objective. While stability is a basic human attitude the greater the success of an organisation the greater the pressures to change.
- Continuity - In circumstances where there are no significant internal or external pressures for change and no need to 'do better', the objective may be to maintain the current direction. This is more typically an objective of social or governmental organisations.

It is quite possible for no one type of objective to be dominant at anygiven point in time or for the primary objective to change, often frequently, as external pressures change and the circumstances of an organisation alter.

All organisations will have a number of financial objectives which, like strategic objectives, are constantly changing to meet circumstances. These form an integral part of the management of the organisation and are closely tied to the strategic objectives. In business financial objectives have always been of prime importance, but there is a trend in many countries to make social and government organisations equally aware of the financial constraints upon them by requiring them to implement stringent financial planning. This is viewed by some as being unnecessarily restrictive on the organisations, since financial considerations often take precedence over social needs. In the GIS community, the more rigorous cost recovery policy imposed by the UK government on the Ordnance Survey is a case in point. This is an example discussed further in Chapter 7.

Closely related to the financial targets are the resource objectives. These can include such aspects of the business as human resources, including how many people are employed, at what levels and with what skills; physical resources such as infrastructure, buildings and plant; and information resources, which for many organisations constitute their most valuable asset and yet one which is too often undervalued and poorly maintained. Balancing the resource objectives with the financial objectives is often the key to achieving the overall strategic objective.

Individuals will also have their own objectives, which may be in conflict with those of the organisation. These may relate to pay, job satisfaction, career structure and so on, and like other objectives will change as circumstances change. Part of the management task is staff motivation, which is affected by individual's perception of how well their personal objectives are being met.

How objectives are set is determined by the culture of the organisation, the experience and attitudes of those making the decisions, prevailing external conditions and any forecast about how these might change, and the availability of information on the status of the organisation. Successful objective setting can only realistically be achieved after information gathering mechanisms have been put in place with appropriate levels of consistency and methods of aggregation. It is too easy to set objectives that do not relate to prevailing conditions at a lower level in the organisational structure.

1.5 The management of information

The management of an organisation can be grouped into four functional areas:

- The effective use of resources

- Planning of future activities
- Improvement of the organisation's overall effectiveness
- Identifying and evaluating the effects of external factors

Each of these requires information either from within the organisation or derived externally. The success of the management functions can only be achieved by the collection of the right information in the right way. Facts about current events may not be enough except where immediate decisions are being made in response to unexpected problems. More often the decisions of a manager are based on information that has accumulated over a period of time. Knowledge about the way particular factors have changed and how these changes may have influenced the processes of the organisation is a vital part of making decisions. In all probability a decision will have effects some way into the future.

The management of information is extremely important. How records are kept, the way in which they are maintained, the accuracy and currency of the information at any given point in time, the sources used and the criteria adopted in collecting the information are all relevant when trying to evaluate such data. It is for this purpose that most organisations keep complex information systems about their most important functions and resources, such as financial control systems, personnel records, resource inventories, product lines and so on. This aspect of information management is extremely important in the handling and successful use of spatial data. It is a theme to which we return in several chapters of this book when considering the role of spatial information in the decision-making environment and the way in which results are evaluated.

The collection of raw data is important but is not in itself enough. There are various levels of data interpretation, particularly where decisions are being made as part of a planning or forecasting process. Interpretation of historical data and judgements about the nature of the relationships between datasets is required. In many cases mathematical modelling is a necessary component of this interpretation in order to identify trends and calculate probabilities.

The importance of being able to ask 'what if?' type questions is one of the fundamental qualities of all information systems and has been stressed as a being a key concept for GIS (Maguire, 1991). It provides the decision maker with tools for deriving a variety of solutions to problems by adjusting the component information elements and their various interactions. The ability to do this quickly and efficiently for spatial data should be a prime motivation for investing in GIS technology.

Such information management presupposes that the information structure is flexible enough to incorporate data that can be manipulated in this way and, importantly, the user understands the implications inherent in changing the data. It is easy to suppose that all solutions proposed by an information system as part of a series of 'what if?' scenarios have equal validity. This is not necessarily the case and judgements will have to be made on the relative merits of each.

The problem of siting a road bypass for a small rural town in England is a case in point. While the consensus of the general public was for a bypass to alleviate the congestion in the high street, none of the various solutions proposed gained widespread public support. In one solution a historical landmark was in danger of being destroyed, in another the cost was a much more critical factor, and in a third, the loss of properties and high quality farm land was deemed unacceptable. The final decision has to weigh the various alternatives carefully, address the relative importance of the different factors and apply considerable judgement in order achieve the result which best meets the objectives of the various parties involved.

1.6 Information networks

Within an organisation management is made more complex by the flow of information. There is a link between decisions and actions which result in information passing up and down and across the organisational structure. This flow of information and the patterns created can be viewed as a model by which organisations may be identified. Modelling these flows is an important part of understanding the decision-making structure of an organisation and is a necessary prerequisite in implementing information handling systems.

In essence information flow has input and output components, elements of storage, transmission and processing and there are rules and external factors which govern these processes. This information flow can be the key to understanding why organisations do not achieve their objectives. As organisations grow and the information system becomes more complex the need to implement computer-based information systems grows. This is particularly so for spatial information as more users identify their need to make greater use of the spatial context of their information resource. One of the main reasons why GIS implementations have failed is their lack of integration within the decision-making structure.

Passing information to other users within an organisation may result in it being altered in some way. Raw data may be selectively interpreted at more than one stage, data may be aggregated in different ways for different purposes and different users may keep their own subsets of the data. This may result in a very uncontrolled management of information with old datasets often being used in preference to the most current versions or only part of a dataset being used to make a decision because there is no knowledge that a fuller dataset exists.

This is a particular problem for map based data and the information associated with it. Many people are reluctant to replace dated maps, witness the number of motorists who keep very old atlases in their cars, and yet the same people demand current telephone directories. An example of the potential for error is a county council in the United Kingdom which between its various departments had not only the centrally maintained map collection with complete and current coverage of the basic topographic map scale of 1:50,000, but also a further 6 separately maintained coverages at that same scale, all of varying currency. In addition, one department also maintained its records on the old topographic map series at the scale of one inch to the mile (1:63,360), which was replaced by the 1:50,000 series in the late 1960s. Each of these map sets was used as a base for additional information and decisions were being made across departments using different basic reference sources.

A further concern within an organisation is the problem of information security. Some information may be regarded as sensitive and only appropriate for certain individuals. This may be for various reasons, but it may properly be the case that information is not released to those who are not in a position to interpret it correctly. Often information is controlled by letting a large number of individuals have access to read the information but no ability to copy or amend it. Quite complex controls may be placed on the flow of information which can be both permanent and temporary.

1.7 Information systems

As a consequence of the complexities of the decision-making process and the accompanying information management problems, there is the need to establish rigorous information systems. These may be manual but more typically are computer based because of the volumes and speed of processing demanded.

A system may be defined as a group or pattern of associated activities which will normally have the following elements (Anderson, 1986):

- A common purpose
- An identifiable objective
- An established sequence of procedures and data flows with at

least one but possibly many elements of input, movement, action, storage and output
- Feedback of information, giving control over the system
- A boundary that defines the extent of the system
- Dependence on specific data

Figure 1.2 represents these elements graphically and shows that there may be both one-way and two-way movement between elements and that inputs and outputs may be both internal to the system and external, in other words associated with other systems. Those systems which contain elements of input from an external source are described as open, while those in which all the elements are internal to the system are described as closed. Information systems and organisation systems are essentially open since they are affected by a variety of external influences, although for the purposes of modelling an organisation it is often easier to identify systems as essentially closed.

It is of course overly simplistic to describe an organisation as a single system. At any one level a system may be sub-divided into a number of sub-systems. Each of these sub-systems may, at a more detailed level, be viewed as a system in its own right and may also contain its own sub-systems. Furthermore there is no simple hierachical

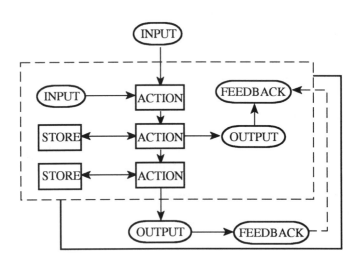

Figure 1.2 Elements of a system (based on Anderson, 1986).

structure to the systems in an organisation. Different systems exist which will overlap, use common inputs and possibly generate common outputs. The systems will be inter-connected by the flows of information and the various control mechanisms established.

This raises a particular problem that needs to be addressed when defining the systems within an organisation - that of system boundary. The boundary not only delimits the extent of a system but also defines the interface between two or more systems at points of input/output. In some places the boundary of a system may be viewed as 'hard'. In such cases the interactions are well defined. For example, system A in Figure 1.3 is responsible for completing form 1 and passing it into system B. System B cannot operate without the form completed in system A.

On the other hand the boundary to a system may be 'soft' where a common set of procedures or actions is shared by more than one system. System A may normally complete form 1 but if it does not system B may access the necessary information and complete it. The sub-system of completing form 1 is therefore shared by both systems A and B (Figure 1.3). Walsham (1993) has stressed the need to understand the context of the information system and to understand the processes whereby the system influences and is influenced by this context.

A further idea that may be introduced at this stage is the notion of 'decision nodes'. These are the points within an system where particular decisions are taken. These decision nodes may be divided into primary and secondary. Primary nodes are those that cannot be bypassed in any way and represent potential 'bottlenecks' in the system. The secondary nodes are those that must be taken but will not necessarily prevent further progress through the system if they are delayed for a time. Thus if staff will only receive their expenses when the claim is signed by the section head, no expenses will be paid when that individual is on holiday. If, however, expense receipts may be submitted for payment with a pre-defined level of retrospective signing the system may continue to partially operate even when the head of section is away. The identification of the decision nodes within a system is an important part of reviewing the efficiency of that system with a view to the installation of a new information system.

When designing information systems the key is the output that is required. Also the structure of the system and how it operates will be of importance since resources will be committed to operating the system. Once a system has been installed it is necessary to regularly monitor it in order to ensure it is operating effectively and achieving the desired goals. It is necessary to identify a number of elements about the system in order to monitor its operation. These include:

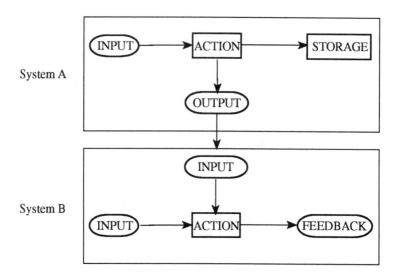

1. Hard system boundary

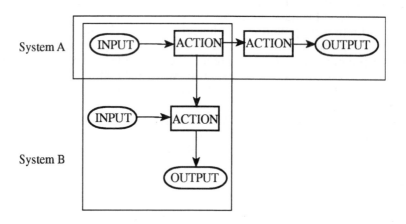

2. Soft system boundary

Figure 1.3 Types of system boundaries.

- Knowing what the system objectives are;
- Defining the information that is found within it;
- Knowing the volumes of data involved;
- The form of the various inputs and outputs;
- Any prescribed ways in which data must be stored;
- What operating and control procedures there need to be;
- What is the timing and frequency of the processes.

All these are important for spatially-based information systems but often key elements have been ignored or incorrectly estimated, such as volumes of data and the processing times involved to transform between one format and another.

1.8 Geo-information systems

The concept of the system is a key one in understanding the role of spatial information management. Appreciating the complexity of the interactions between systems within an organisation is important for spatial information managers who are required to design and operate systems that, in part at least, utilise spatial data. The introduction of GIS to facilitate greater efficiency and to provide further information handling capabilities in the use of spatial data requires a clear understanding of the operating environment.

The failure to define clear and achievable objectives for the use of spatial data can lead to the development of inappropriate applications using GIS technology. The objectives of the information system may not explicitly recognise the need to utilise spatial data to achieve a particular output, since a number of related processes may be undertaken, one or more of which involves some spatial query or manipulation.

A local authority may maintain a record of road accidents and will use this information to identify areas prone to particularly types of incident, often called accident black spots. Once identified this can lead to remedial measures being undertaken. Such a system will typically collect information on the nature of accidents from police reports which give details about those involved, the types of vehicles, time of day, weather conditions and any other contributing factors. The reports will also contain details of the road layout, final positions of vehicles and any measurements taken at the scene such as skid marks. From these inputs a database is created, which in part is spatial, since the information relating to the final position of the vehicles is geocoded. This is shown graphically in Figure 1.4.

The criteria for classifying an accident black spot will consider a

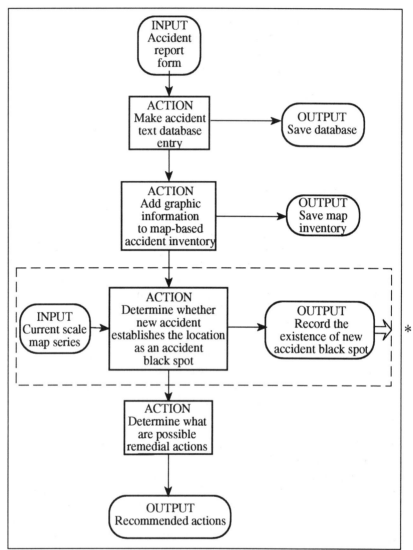

* Inventory accessed by other systems

Figure 1.4 Accident black spot information system.

number of variables, including defining a location where a number of similar accidents have occurred. How the location is determined may vary depending on the size of junction and nature of the accidents. A number of accidents involving children during the same period of time each day, 3.30pm to 5.00pm, on a short stretch of road adjacent to a school may be easy to identify and the remedial action be the placement of a road crossing facility.

On the other hand a number of multi-vehicle accidents may have occurred along a road which has a long and steep gradient. The cause of the accidents may be driving too fast but the accident distribution along the road may make the location of a black 'spot' hard to identify. Alternatively the information system may be used to identify a general increase in accidents at night without any consideration of spatial context, with the resulting remedial action being a driver-awareness campaign on the use of headlights.

To achieve the objectives of this type of information system elements of spatial query and manipulation may be used, such as 'the number of accidents within a specified distance of point X'. On the other hand the use of spatial information is not a pre-requisite to achieving the system objectives and the role of spatial data is one of a potential input in the same way that information on finances or physical resources may be included in determining future actions.

For many systems the collection, storage and use of spatial information is an integral part. Spatial data may form an underlying framework from which relative and actual location may be determined. Those systems which use a spatial referencing system as the basis for information can be called geo-dependent information systems. The following distinction may be made:

- Geo-dependent information systems: Those systems which require the use of some element of spatial data to achieve their objectives.
- Non-geo-dependent information systems: Those systems that may utilise spatial data to achieve their objectives but where it is not a prerequisite for the system to operate.

Non-spatial data will almost always be used in conjunction with the spatial information framework, thus there is the need to ensure a high level of information integration. This may occur at an actual level within a single database or a number of linked databases, or may occur at a conceptual level within the system which allows information of different types and in different forms to be accessed and interpreted. Issues relating to the nature of information integration and the importance of the spatial context are discussed in Chapter 4 when considering integrated approaches to geo-information management.

Spatial data are complex and expensive to generate, both in time and resources, and one dataset will often be used by more than one system. It is possible to identify several non-spatial information systems which share the same geo-information subsystem. For example the map-based accident inventory may be used not only to determine accident black spots but by the police to optimise resource allocation to apprehend those committing traffic offences.

The introduction of GIS technology to support information handling has to recognise the complexity of system and subsystem interaction. The potential to improve the quality of information handling with GIS technology will often only be achieved by changing the nature of a system's processes, the form of its inputs and outputs and, in some cases, redefining the system objectives. The outcome of this process can be organisational change and restructuring which is not always an acceptable or desirable outcome of technical innovation. The case for integrated implementation of the technology has to be made on the basis of careful definition of objectives at the organisational and system levels.

Many GIS solutions have been limited in scope with a tendancy not to be integrated within the broader information strategy of an organisation. Information integration is the key to successful geo-information management. In some cases the ability to handle spatial information can lead to a complete review of the information handling environment and a reassessment of the criteria for decision-making.

1.9 Spatial decision support systems

While the information system has to reflect the availability and use of spatial data, so the decision-making culture has to adjust to include the results of spatial information manipulation and analysis. To aid the decision maker in deriving acceptable solutions, computer-based tools have been developed that can evaluate large volumes of data and utilise modelling functions to help identify relationships and trends within the data. These tools include decision support systems (DSS) which are used to support operational research in finding solutions to business problems and executive support systems (ESS) which provide senior managers with information linked to specific business objectives (Laudon and Laudon, 1988; Rockart and DeLong, 1988).

Silver (1991) defines a DSS as a 'computer-based information system that supports people engaged in decision-making activities'. The 'support' component refers to the assistance the system gives the decision maker in exercising judgement; the system itself does not make the decision. DSS were developed as a result of the shortcomings of management information systems (MIS) in the late 1960s and early 1970s which did not adequately involve the decision maker in the process of finding the appropriate solution. They are defined in terms of the characteristics of the system. Geoffrion (1983) suggests that a DSS has six distinguishing characteristics (Densham, 1991):

- They are explicitly designed to solve ill-structured problems where the objectives of the decision maker and the problem itself cannot be fully or precisely defined.
- They have a user interface that is both powerful and easy to use.
- Such systems enable the user to combine analytical models and data in a flexible manner.
- They help the user explore the solution space (the options available) by using the models in the system to generate a series of feasible alternatives.
- They support a variety of decision-making styles and are easily adapted to provide new capabilities as the needs of the user evolve.
- DSS allow problem solving to be both interactive and recursive - a process in which decision making proceeds by multiple paths, perhaps involving different routes, rather than a single linear path.

From these it is clear that DSS has a role where complex information systems involving various inputs, outputs, actions and feedback require an evaluation of the possible solutions and their implications at key decision nodes in the information flow. Many problems are ill-structured, that is to say, it is not clear what are the boundaries to the problem under consideration and what are the objectives of a particular decision maker. A more comprehensive review of DSS can be found in Bidgoli (1989).

Various authors have identified the need to develop the DSS concept with respect to spatial information (Armstrong *et al.*, 1986; Densham and Rushton, 1988; Densham and Goodchild, 1989; Armstrong and Densham, 1990; Densham, 1991). If, as has already been noted, it is implicit that GIS are designed to support spatial decision making then, as Densham (1991) notes, there are limitations with current GIS solutions which prevent it being readily used as a DSS tool. These are essentially as follows:

- The lack of appropriate analysis capabilities;
- The nature of GIS databases and the inherent design limitations that basically only support cartographic display;
- The limitation of information communication to a cartographic

form and simple tabular database output;
- The lack of flexibility in the way a GIS approaches the whole process of spatial decision making.

The development of spatial decision support systems (SDSS) is an attempt to bring together the decision-making benefits to be gained from DSS with the basic spatial information handling capabilities inherent in a GIS. Thus an SDSS will not only have the above characteristics which distinguish a DSS but will also need to provide additional capabilities and functions (Densham, 1991):

- Provide mechanisms for the input of spatial data.
- Allow representation of the complex spatial relations and structures that are common in spatial data.
- Include analytical techniques that are unique to both spatial and geographical analysis, including statistics.
- Provide output in a variety of spatial forms including maps and other, more specialised types.

An SDSS is characterised by using spatial data as part of the problem solving process. This process involves a high level of participation by the decision maker and is essentially an iterative one in which the decision maker constantly refines the solution according to his perceived objectives and expert knowledge. Another important aspect of a SDSS is the integration of the spatial and non-spatial data, providing the widest possible information base on which to develop appropriate solutions. Another important element of an SDSS is the use of models to generate solutions. The existence of a component which manages the creation and enhancement of suitable models adds a further component not available in contemporary GIS solutions.

A simple structure for an SDSS is shown in Figure 1.5. This shows the important information system elements and the need for an integrated software solution between the various components which is relatively transparent to the user due to the power of the user interface. It also shows the two-way interaction with the decision-maker.

An example of a DSS developed for spatial data is the geodata analysis and display system (GADS) described by Silver (1991) after Sprague and Carlson (1982). The purpose of this system was 'to improve the effectiveness of professionals by giving them access to useful, graphically-oriented, computer-based facilities, but not to impose any particular decision-making process on them' (Silver, 1991). Studies showed that processes changed and decisions improved with the use of GADS, even though individual decision makers varied in terms of how they used the system to arrive at decisions.

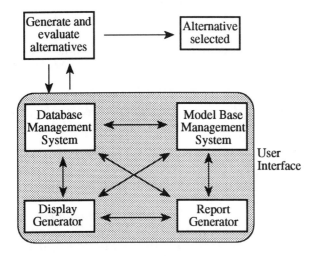

Figure 1.5 A structure for a spatial decision support system (Armstrong, Densham and Rushton, 1986).

References

Anderson, R. (1986) *Management, Information Systems and Computers. An Introduction.* Macmillan, London.

Aranoff, S. (1989) *Geographic Information Systems. A management perspective.* WDL Publications, Ottowa

Armstrong, M. P. and Densham, P. J. (1990) Database organization alternatives for spatial decision support systems. *International Journal of Geographical Information Systems* Vol 4, No 1, pp. 3-20.

Armstrong, M. P., Densham, P. J. and Rushton, G. (1986) Architecture for a microcomputer-based decision support system. *Proceedings of the 2nd International Symposium on Spatial Data Handling.* International Geographical Union, Williamsville, pp. 120-31.

Bidgoli, H. (1989) *Decision Support Systems, Principles and Practice.* West Publishing.

Burroughs, P. A. (1986) *Principles of Geographical Information Systems for Land Resources Assessments.* Clarendon Press, Oxford.

Densham, P. J. (1991) Spatial decision support systems In:

Geographical Information Systems Volume 1: Principles. Eds. Maguire, D. J., Goodchild, M. F. and Rhind, D. W., Longman, London, pp. 403-12.

Densham, P. J. and Goodchild, M. F. (1989) Spatial decision support systems: a research agenda. *Proceedings of GIS/LIS '89.* ACSM, Bethesda, pp. 707-16.

Densham, P. J. and Rushton, G. (1988) Decision support systems for local planning. In: *Behavioural Modelling in Geography and Planning.* Eds. Golledge, R. and Timmermans, H., Croom-Helm,   London pp. 56-90.

Enyon, D. (Ed.) (1992) *1993 International GIS Sourcebook,* GIS World, Fort Collins.

Geoffrion, A. M. (1983) Can OR/MS evolve fast enough? *Interfaces* Vol 13, pp. 10-25.

Laudon, K. C. and Laudon, J. P. (1988) *Management Information Systems. A Contemporary Perspective.* Macmillan, New York.

Laurini, R. and Thompson, D. (1992) *Fundamentals of Spatial Information Systems.* Academic Press, London.

Maguire, D. J. (1991) An overview and definitions of GIS. In: *Geographical Information Systems Volume 1: Principles.* Eds. Maguire, D. J., Goodchild, M. F. and Rhind, D. W., Longman, London, pp. 3-7.

Rockart, J. F. and DeLong, D. W. (1988) *Executive Support Systems. The Emergence of Top Management Computer Use.* Dow Jones-Irwin, Illinois.

Silver, M. S. (1991) *Systems That Support Decision Makers. Description and Analysis.* Wiley, New York.

Sprague, R. H. and Carlson, E. D. (1982) *Building Effective Decision Support Systems.* Prentice Hall, New Jersey.

Walsham, G. (1993) *Interpreting information systems in organizations.* Wiley, New York.

2

Data for geo-information systems

2.1 Data into information

Before considering the types of data used in spatial information systems it is necessary to clarify the distinction between data and information. We may for the purposes of this discussion refer to data as being raw or unprocessed, while information has undergone some processing, such as classification, which makes it more relevant to the problem in hand. Thus data are the original survey information, the original remote sensing image or the basic census statistics, while information is the cartographic representation of the survey data, the classified remote sensing image or the aggregated census statistics. The process of converting from data to information is one which 'adds value' due to the 'knowledge' required. For example, the classification of the remote sensing image is only successfully undertaken by someone with the knowledge to apply the necessary statistical techniques.

At different levels within a decision-making hierarchy, interpretation or analysis is undertaken, so adding extra value to the original data. In a sense within a single process 'data' are taken in and information is produced. So there is a hierarchy of levels in which basic data are progressively converted into more and more complex information. It is possible that the knowledge added or the processes undertaken are inappropriate or inadequate and thus the information is of less value than the original data. It is therefore important to identify what is data and what is information within the context of the decision-making process.

The conversion of data to information raises a number of issues which have for some time been under discussion in the GIS community. One problem is that of data ownership and who has the copyright on information which has added value given to it by someone other than the original data collector. A second is the value of information which does not wear out but which may or may not diminish in value through time. Certainly a geology map from the late 1800s may be as valuable today as when it was first interpreted, but census data may lose a lot of their value once they are superseded by the next census. Copyright, ownership and some of the legal aspects of GIS are discussed more fully in Chapter 6.

2.2 Describing geographical location

Many of the problems faced on a regular basis utilise an understanding of geography to help solve them, either implicitly or explicitly. The way features are recognised and grouped is important to our perception of geographical space. Some locations are perceived in very clear terms, such as the extent of the property which we own or rent. Other features such as our 'home town' will have a geographical extent which we recognise but which may or may not accord with representations such as the administrative and political boundaries. Other geographical areas such as the Alps have rather vague regional connotations but no precise geographical extent in most people's minds.

Some elements of geographical location are defined in terms of adjacency or proximity. You may not know in precise terms where Church Road is but in directing a stranger it is easy to identify that it is the turning 'next to the church'. The village of Shalford in England would normally be described as being 'near Guildford' in order that people may better understand its location.

Table 2.1 after Raper *et al.* (1992) gives the characteristics of different methods for describing geographical location. It can be seen that understanding how we describe location is important to the type of data we collect, how we structure spatial databases and how we undertake information retrieval.

2.3 Objects and hierarchies

Spatial decision making is concerned with objects and generic groups of objects and not cartographic entities. The highway engineer is concerned with the object 'road' or the generic grouping 'road network', which may be defined in various ways. He is not necessarily concerned with the lines and symbols that are used to represent a road on a map. Figure 2.1 shows three levels of cartographic representation for part of a road network at different scales. In each map there are specific cartographic entities such as lines, symbols or text. There is also a hierarchy of object definition from the basic cartographic entity which may be the road kerb, to the object 'road' and the generic name for a series of linked roads, the 'road network'.

Type	Characteristics	Example	Advantages	Disadvantages
Nominal	No relationship between 'objects' can be derived	County name, individual name of house	Simple, familiar	Useless except for distinguishing one place from another. Difficult to handle inside computers because it varies greatly in form
Partially sequenced	Order of house numbers normally in sequence (2,4,6 or 1,3,5) within a street, i.e. some geographical hierarchy implied	Street address	Simple, familiar and matches to structure of the real world. Widely used in manual record keeping	Relationship between house numbers approximate, that between streets unknown
Topological	Defines what is next in space to a particular 'object'	'Oxford Street and Regent Street intersect' or 'Kent is adjacent to Sussex'	Familiar and simple in terms of route following	Does not imply physical distance; normally, maps cannot be made on this basis
Local geometry	Defines what is next in space to a particular object and quantifies the distance between objects	'The hydrant is 5 metres from the corner of this house'	Easy to understand and use, works well over short distances and where landmarks are common	Not good in rural areas with few landmarks or over large areas. Cannot relate objects described in relation to different landmarks
Global geometry	Defines location in space as compared to a single fixed point (e.g. origin of the British National Grid or the centre of the Earth)	'At National Grid Reference NT123 456' or 'at latitude 51 North and longitude 1 West'	Works well over large areas; permits calculation of distance between any two objects. Can be used at any level of detail. Simple to handle in computers - standard in form	Not 'natural' on the ground; requires maps to identify location or satellite-based position fixing systems

Table 2.1 The characteristics of the different methods for describing geographical location (from Raper *et al.*, 1992)

How an individual defines 'road' will vary, depending on the nature of the problem being addressed and the scale at which the solution is to be found. On a large scale plan a road may consist of several cartographic entities which define the road kerbs, pavements, road centre line and road features such as drains and manhole covers, road markings and street furniture. In trying to identify the road along which a delivery lorry is to be routed, details about the pavements and street furniture are not important. Instead the links between elements in the road network are vital as well as particular constraints on the use of a road such as width restrictions, speed limits and one-way traffic flow. On the other hand the design of a new pedestrian crossing will utilise a lot more detailed information about the road surface, width, pavements, visibility of the crossing and so on.

Thus any data model must take account of both the different cartographic entities and the spatial objects represented at various levels of representation. The ability to identify spatial objects which are composed of a number of cartographic entities together with other attribute data takes the GIS away from being a map management tool to being a spatial information management tool. This requires a clear understanding of the types of spatial data and the limitations inherent in the way spatial information has been collected and how it is currently structured.

2.4 Maps and cartographic representation

The variety and nature of spatially-referenced information represents a large information handling problem. The traditional method of representing spatial data is the map. The map is a medium for the comprehension, recording and communication of spatial relationships and forms. It is an abstract model of reality which includes transformations of various kinds and conveys, directly or implicitly, various sorts of information, such as location, direction, distance, height, connectivity, contiguity, adjacency, hierarchy and spatial association (Visvalingham, 1990; Cassettari *et al,* 1992).

Maps contain different levels of information representation depending on the type of geographic phenomena shown. Some features can be correctly portrayed using linework, others are simplified to symbols or text depending on the scale of representation, the information theme of the map and the communication objectives. In a number of European countries the scale of maps produced by the national mapping agencies range from small scale thematic maps to large scale maps at anything from 1:5000 to 1:1250. The purposes of these maps range from general information for national planning to a

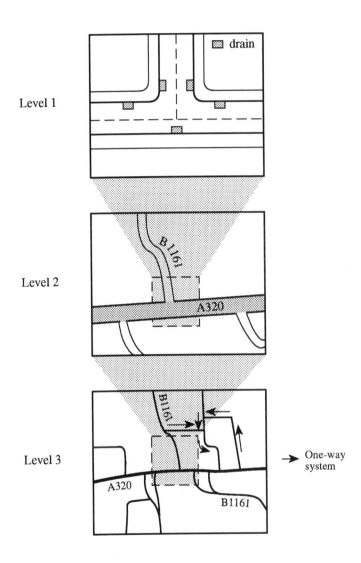

Figure 2.1 Cartographic and object hierarchy showing the changes in feature representation as a result of scale.

detailed property inventory showing building outlines and features that mark land ownership boundaries. Further to the basic topographic and large scale plans are thematic maps showing specific types of information such as geological data, soils information, hydrological features, social information and so on. Add to these air charts and hydrographic charts for air and sea navigation, the map represents a comprehensive information source of the three dimensional environment.

The map can be the most detailed form of spatial information representation. Dangermond (1979) classified the wide range of geographic data types according to their method of presentation in a cartographic form (Figure 2.2). This is based on the premise that all elements in a map may be broken down into points, lines or areas. The consequence of such a wide range of geographic representations is that the map requires appropriate interpretation skills to be fully utilised. Those who are familiar with relief representation on a topographic map by the use of contours will be able to convert the two dimensional line image to a three dimensional mental model of the terrain using the key symbology, their own experience of topographic representation on maps and their background knowledge of an area. An experienced hill walker will, for example, be able to determine the steepness of the slope from the contour interval, the scale of the map and the contour spacing. But there are also many occasions when misinterpretations of maps has led to fatal or near fatal incidents.

The map is also a reduced representation of reality and not just a scaled down image. The map as a communication medium involves the use of processes such as selection, classification, displacement, symbolisation and graphic exaggeration (Visvalingham, 1990). There is an important distinction between data held in a digital geographic database and the data presented in map form. This is one of the major problems for the effective use of GIS, since spatial analysis requires accurate spatial location while the effective communication of analysis results requires the use of cartographic principles such as generalisation and cartographic design.

2.5 Considerations in using maps

If the map is to be integrated within a broad decision making environment, system designers have to understand the potential limitations of such information. These were summarised by Rhind and Clark (1988), and may be categorised as follows:

- *Scale*: The scale at which an analogue map is compiled

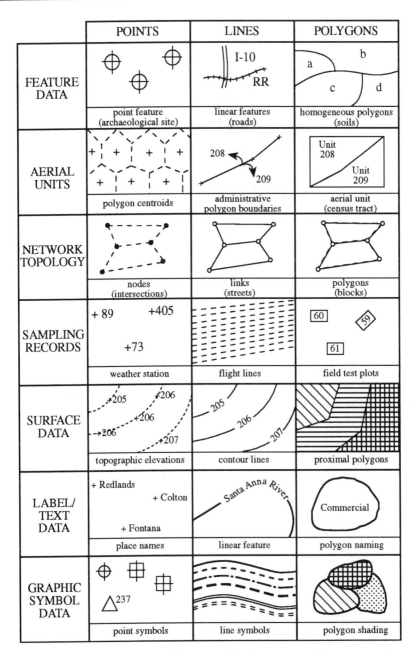

	POINTS	LINES	POLYGONS
FEATURE DATA	point feature (archaeological site)	linear features (roads)	homogeneous polygons (soils)
AERIAL UNITS	polygon centroids	administrative polygon boundaries	aerial unit (census tract)
NETWORK TOPOLOGY	nodes (intersections)	links (streets)	polygons (blocks)
SAMPLING RECORDS	weather station	flight lines	field test plots
SURFACE DATA	topographic elevations	contour lines	proximal polygons
LABEL/ TEXT DATA	place names	linear feature	polygon naming
GRAPHIC SYMBOL DATA	point symbols	line symbols	polygon shading

Figure 2.2 Classification of geographic data types (Dangermond, 1979).

determines the smallest area that can be drawn and recognised on a paper map. The eye can resolve to about 0.2mm or a resolution of 10 metres at a scale of 1:50,000, but at that scale it is not possible to represent accurately any object of less than about 25 metres across (Fisher, 1991). Where analogue maps are converted to digital databases the scale limitations inherent in the original map are translated to the database. This is important in understanding the accuracy of objects portrayed in the database and appreciating what objects have been included and what excluded. A digital database compiled from a number of maps at different scales will contain various levels of information representation and accuracy. It is important to note that scale in itself is not a measure of accuracy.

- *Accuracy*: The required spatial accuracy acceptable for a particular problem has to be specified and understood. Accuracy can be divided into positional accuracy, measurable in all three dimensions, and the classification accuracy: for example, is the lane marked on the map actually a tarmac road or is it a dirt track? Different problems will require different orders of accuracy. A utility will need to know the position of its underground pipes to the nearest metre in order to locate them when carrying out repair work, while the environmental manager will only define habitat boundaries in generalised terms. A number of map accuracy techniques have been developed in order to establish cartographic standards for maps, for example the use of root-mean-square error terms for planimetric co-ordinate accuracy (Fisher, 1991). These have been used, along with qualitative measures of accuracy, to define standards for national map series. The problem faced by the user of digital map data are to reconcile the accuracy of cartographic objects derived from different scales of map and different types of map; for example Dale and McLaughlin (1988) note that some of the least accurate of data sources are the records of legal property boundaries.

- *Currency*: Analogue maps are effectively a snap shot in time of the spatial entities as they existed when they were surveyed. Different maps in a single map series will be of different dates, leading to inconsistencies of information representation at the map edges. In some areas of the world medium scale topographic maps have never been produced or date back thirty years or more. The value of such information will depend on the type and magnitude of change. There are also applications where historical spatial information is required.

- *Cartographic modelling*: A lot of valuable information may be derived from a map by the use of measurement and modelling techniques. Manual methods of analysing maps are often repetitive and take a long time. There is the potential for automated techniques to be viable in an integrated decision making context. Without these techniques many map based information systems are purely information inventories that are visually interpreted.

- *Visualisation.* Maps are very striking graphics which use colour, symbols and text in imaginative and creative ways. The development of a map specification and the compilation of the map itself is in part dictated by the potential users and how the creator wishes the maps to be interpreted. Poor design can lead to poor or inaccurate reading of the map by the user.

2.6 Map-based geo-information systems

The map provides a means by which specific problem-oriented information may be recorded in the correct spatial context. Annotated map overlays and extractions from analogue maps are produced as part of a decision making process for a wide range of tasks, from urban planning and the design of new roads to environmental management and the monitoring of social change. In many organisations the map has become an essential part of the information resource, where it is typically used by specialists who have an understanding of the underlying cartographic principles.

The relationship between cartography and GIS is not altogether clear at the practical level. For some, GIS is still regarded as a visually interpreted map-based inventory. On the other hand, where a GIS is used for its analytical capabilities in the integration of spatial and non-spatial databases, cartography forms the basis for the display of information. Even for those users of GIS who do not see the visual map as an essential component, cartography has a role in the creation of the digital map base. Cartographic principles are important in the wider spatial information handling context and need to be viewed as central to understanding the nature of spatial data. Figure 2.3 shows the nature of the relationship between cartography and non-spatial databases in a GIS framework. The development of GIS is creating an awareness of cartography as a subject in its own right. In this context the map is the communication medium for geo-information.

Any information system in which maps play an important part in the decision-making process must be regarded as a geo-information system, since the spatial context is an integral part of the whole process. As an example, the monitoring of planning applications through local

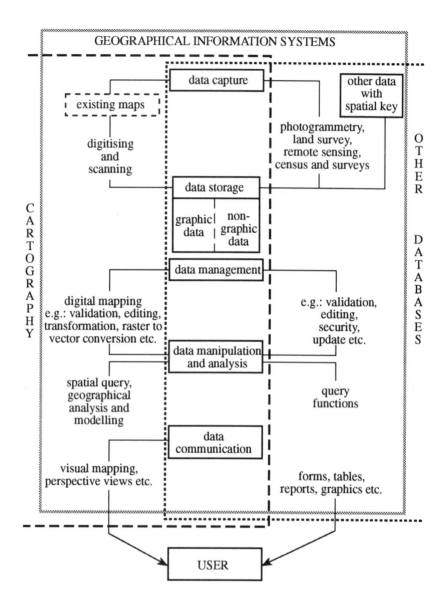

Figure 2.3 Relationship between cartography and non-spatial databases in a GIS framework (Cassettari *et al.*, 1992).

authority planning departments involves the registering of the planning proposal, with the intended changes to building and infrastructure drawn onto base maps. Information relating to the application, whether or not successful, is retained so that future applications or amendments can be considered not only in the light of existing circumstances but with due consideration to the reasons why previous applications have been rejected. Such considerations may be very important if adjacent applications conflict with each other. The use of the map in this context is essential to the process of making a decision about the acceptability of a planning application.

2.7 Data derived from spatial analysis

Presenting information in a cartographic form is a powerful method of visual communication for many decision makers. However, it is not always necessary or desirable to use a map. Information may be derived by analysing data within a spatial framework but presented in a non-cartographic form such as text, tables or graphs. The spatial framework may be global in nature, such as using latitude and longitude, or may be some arbitrary grid which has no other meaning than to present the relative position of local information, such as might be used on a town plan.

Spatial statistics use the spatial framework in which to apply quantitative and statistical methods, including mathematical modelling and operational research methods, for a better understanding of spatial relationships. Spatial analysis based on maps fall into two categories: spatial pattern description and spatial pattern relationships.

There are problems in implementing spatial analysis techniques in GIS. Current systems do not have the tools for identifying and describing spatial patterns and do not provide for hypothesis testing. The lack of a modelling capability has already been noted in connection with requirements of spatial decision support systems (Chapter 1).

The results from such spatial analysis should form an important output from a spatial information system. Monmonier (1977) observed that one of the basic problems when using analogue maps for decision making was that they are not easy to use for spatial analysis. In essence the patterns on a map are too complex for easy recognition, and the human brain is easily tricked into identifying patterns that have no basis from the available evidence. There is then a need to recognise when the map is not the right medium for decision makers and for choosing an appropriate alternative form of communication. There is also a need to identify when a more simplified form of the map should be generated from the basic database for presentation of results. This is an issue discussed further in Chapter 10.

2.8 Address-based spatial information

Spatial data does not have to be derived from a map. Much spatial data exists as part of datasets created for use in a non-spatial information system. The address is one example which represents a point or area on a map. The UK's Ordnance Survey, in recognition of the importance of the address as a spatial identifier, produce a product called Address Point. This is derived from the National Address Gazetteer and each postal address in Great Britain (which excludes Northern Ireland) is linked to a one metre resolution National Grid reference, as shown in Figure 2.4. The dataset is generated automatically using Ordnance Survey digital map data and data from the Royal Mail's Postal Address File (PAF). This project demonstrates the potential for linking existing information from maps and other sources, resulting in a derived data set which has 'added value'.

The importance of developing this type of derived dataset is the potential for generating complex spatial relationships from multiple datasets. For example the address data could be linked to road centre line information for the purposes of determining the address of properties affected by road works. It could also be used to develop a unique object reference which refers to ownership of a property or part of a property, and from this to derive unique property reference numbers. The depiction of property ownership presents a problem using traditional maps, since they do not consider the three dimensional nature of buildings and therefore it is difficult to show ownership of flats or apartments in a block or the existence of a residential property above a retail property. The address is one way of accessing this type of spatial relationship.

2.9 Postcodes

The postcode or zip code forms another method of providing a link between spatial location and other information. In essence the postcode is an aggregation of a number of postal addresses and can be used to define areas. The importance of this type of spatial reference is the hierarchical nature of the reference system and thus the ability to analyse data at differing spatial resolutions dependent on the problem being addressed. As an example, the hierarchical zip code system for the United States is shown in Figure 2.5. The UK postcode system and examples from other European countries are detailed in Raper et al (1992).

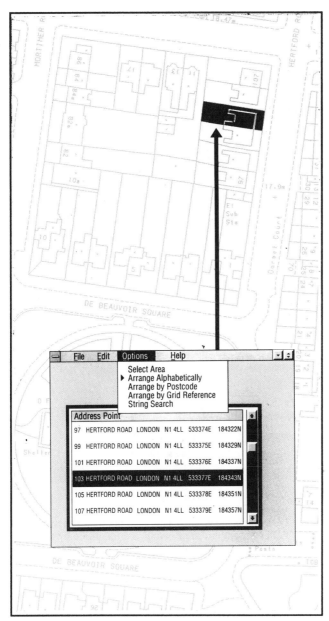

© Crown Copyright

Figure 2.4 Ordnance Survey's Address Point dataset.

ZIP Code National Areas

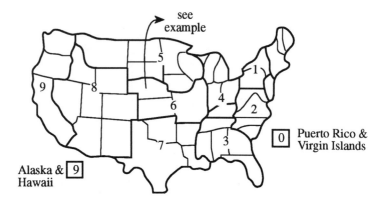

The fisrt digit of a ZIP Code divides the USA into national areas, 10 large groups of states numbered from 0 in the Northeast to 9 in the West.

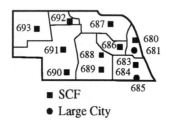

■ SCF

● Large City

Within these karge areas, each state is divided into an average of 10 smaller geographic areas which are identified by the 2nd and 3rd digits of the ZIP Code.

Figure 2.5 The hierarchical zip code system for the United States.

The wide use of postcode information, either linked to address based datasets or used in its own right as a spatial identifier, is valuable because it is a national reference system and because it provides a mechanism for the integration of spatial location and non-spatial data. In the UK linking the postcode with the Address Point dataset and the digital map will provide the opportunity for generating Unit Postcode polygons and accurate Postcode grid reference centroids, as shown in Figure 2.6.

2.10 Other spatial datasets

For many applications, information on the delineation of areas relates to very specific activities, such as conservation areas, military command areas, sales force sectors and so on. Organisations tend to generate their own maps with such boundaries, often in the form of overlays to published maps, using the surveyed information to define the area of interest or to establish the correct spatial context and levels of accuracy.

The Transport Planning Zones Project for the UK Department of Transport supports strategic planning for the road network. A detailed set of zones were established for the London area, compiled by reference to the 1:10,000 mapping. The rest of the south east of England was zoned by reference to 1:250,000 mapping and the remainder of the UK zoned from the 1:625,000 maps. Thus the final digital zone map was derived from a number of scales giving boundary depiction at different resolutions and orders of accuracy. The hierarchical numbering system to some extent reflected the scale at which the zone boundaries had been defined.

A subsequent project determined that the Department of Transport Planning Zones data were inadequate for their purposes. The South East Regional Transport Model (SERTM) project, which focused on the south east of England, reworked the boundary data, simplifying that for the London area by aggregating zones originally extracted from the 1:10,000 map series, and adding detail to the rest of the region by identifying sub-zones from the 1:50,000 mapping. This example is indicative of many datasets being created for use within spatial information systems.

To utilise fully many non-spatial datasets that contain within them a form of spatial identifier it is necessary to create the specific spatial entities to which the dataset refers. The best example is that of socio-economic data collected as part of a national census. As an example, a large amount of time and effort was put into digitizing the boundaries of enumeration districts for the London area so that the local authorities can fully utilise the 1991 census data.

UK Postcode System

BR5 1BX

BR   The Royal Mail area office.
        There are 120.

5   The postal district within the area.
        There are 2700.

1   The sector within the district.

BX   The characters pinpoint the address to within 15 letter
        boxes on average.

Postcode polygons can be created at any level. This map shows
postal district polygons for part of south east London and Kent.

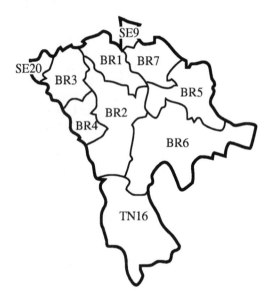

Figure 2.6 Postcode polygons for the United Kingdom.

The defining of spatial entities in this type of dataset is a task that is undertaken at a particular point in time when a certain set of circumstances pertain and specific criteria are used. Thus, as we have seen with the Transport Planning Zones and SERTM data, the resulting datasets may be used in similar types of problem solving, but be quite different in the spatial entities they depict. It is important to understand the history of the dataset and the circumstances leading to its creation before it can be properly used for analysis. This is a problem when comparing the results from more than one census when the boundaries of the units to which the census data are aggregated have changed (Openshaw, 1983). The modifiable areal unit problem (MAUP), discussed by Openshaw (1984), demonstrates that there is a substantial number of ways in which a set of small area units may be aggregated to form larger units and that the resulting statistics can change significantly.

There is also the problem of comparing data collected for one group of spatial entities with data collected on the basis of another group of entities. As an example, the census districts for Surrey, England bear no relation to the Health authority districts. Thus it is very difficult to establish clear relationships between the two groups of spatial entities. The ability to map from one regionalization onto another is also part of the modifiable areal unit problem (see Flowerdew and Green, 1989; Flowerdew, 1991, Openshaw, 1991).

2.11 Attribute data

Spatial attribute data form the other important element in spatial databases. 'Attribute data are complementary to the locational data and describe what is at a point, along a line, or within a polygon.' (Fisher, 1991). At the simplest level they are essential in providing the descriptors necessary to understand the type of entities being depicted by the spatial elements, such as 'road' or 'field boundary' for a line segment, or 'woodland' for a polygon.

At one level attributes are allocated to identify individual spatial elements, such as 'road segment 123' in Figure 2.7. At a more general level, attributes may be used to identify groups of objects such as roads or buildings (Figure 2.7). Hierarchical attribute coding structures can be used to establish complex levels of object definition using single attributes (Figure 2.8). In the more complex data structures some elements have more than one basic attribute since they represent more than one object or object group, such as a line which is part of a road and part of a field boundary (Figure 2.7).

Attributes attached to individual spatial elements may be linked to other datasets in which further detailed information may be stored. In

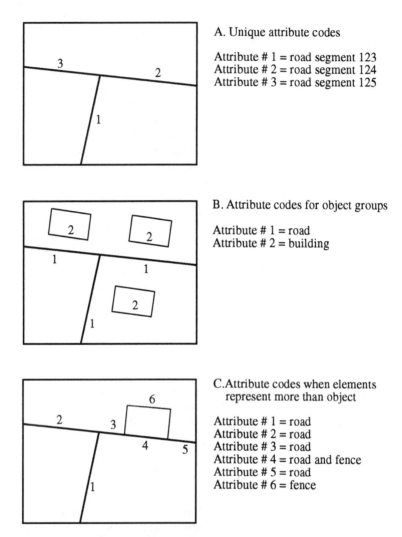

A. Unique attribute codes

Attribute # 1 = road segment 123
Attribute # 2 = road segment 124
Attribute # 3 = road segment 125

B. Attribute codes for object groups

Attribute # 1 = road
Attribute # 2 = building

C. Attribute codes when elements
   represent more than object

Attribute # 1 = road
Attribute # 2 = road
Attribute # 3 = road
Attribute # 4 = road and fence
Attribute # 5 = road
Attribute # 6 = fence

Note that case C is more likely to be handled in a different model,
depending on the data structure.

Figure 2.7 Different approaches to allocating attributes to spatial
entities.

the case of a land parcel and associated house there may be details on the owner, when purchased, price and land area, and house details such as style, number of rooms, facilities and age. Note that this involves the integration of a number of spatial elements combined to create two spatial objects which have a number of attributes, some relate only to the land parcel, others relate only to the house and a third group which relate to both objects (Figure 2.9). So it is possible to have a complex hierarchy of attributes which are directly related to spatial entities.

All these attributes may cross reference to items in other databases and provide a very large amount of extra data which relates to the initial object. These might include details about the owner's credit rating or driving convictions, insurance policies on the house, information from land contamination surveys on the property. To provide the most efficient analysis capabilities and the greatest integration between

Attribute hierarchy

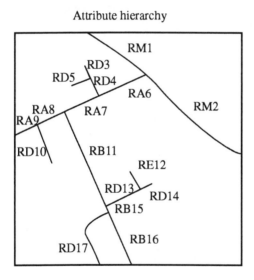

R	Generic object group - road
M/A/B/D	Sub-group e.g. motorway, main
1-17	Individual road segment

Figure 2.8 A simple hierarchical attribute structure for a road network.

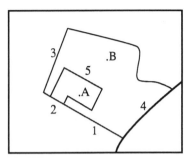

Map of land parcel and house

Spatial Elements	Attribute # 1	Attribute # 2
1	property boundary	fence
2	property boundary	building
3	property boundary	fence
4	property boundary	road
5	building	

Spatial objects	Attribute # 1	Attribute # 2	Attribute # 3
A	house	year built	number of rooms
B	land parcel	area	

Spatial object	Attribute # 1	Attribute # 2	Attribute # 3
A & B	owner	when purchased	mortgage value

Individual groups of

Spatial objects
Spatial objects
Spatial objects

data-base

Figure 2.9 Complex structure of spatial attributes relating to a land parcel and house. Links to other information may be from any level in the structure.

datasets it is necessary to establish an appropriate attribute data model.

2.12 Metadata

With the development of large integrated spatial and non-spatial databases there is a need to keep records about the data. The purpose of these records is to provide background information on the nature of the data, the sources from which it was derived and the overall quality of the material for those intending to use the data for deriving further information. This type of documentary material is referred to as metadata or meta-information.

Cornelius (1991) identified a number of benefits that can be obtained from a better knowledge about the data available for use:

- Prevention of duplication in the collection and storage of data
- Improved access to data
- Standardisation of data formats
- Increased flexibility in using data
- Resultant enhanced value of data resources.

The type of information that may be included in metadata are:

- Sources of spatial information
- Accuracy, quality and currency statements for sources
- Explanations of how source data were used to compile the dataset
- Definitions of attributes and the rules by which they have been coded
- Rules and procedures adopted for data capture
- Results from geometric accuracy tests
- Types of analysis procedures performed on the original data and the constraints and quality of the results
- Rules for the display and cartographic representation of data.

Some of this information may be stored directly as attributes to a spatial entity, such as the source map from which a particular road alignment was extracted. Thus the metadata can be integrated into the spatial information system. On the other hand it may form a distinct and possibly separate dataset. It is possible to have metadata in a spatial form, for example a map may be used to show the scale of source mapping used to compile a digital map, or it may be used to indicate the currency or the accuracy of the information within a digital database (Figure 2.10).

Aerial photographic compilation source index

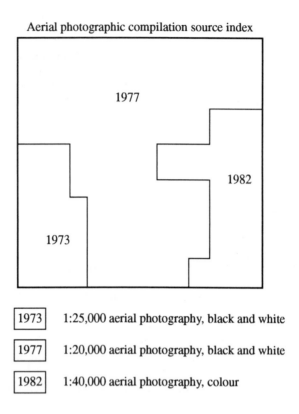

1973	1:25,000 aerial photography, black and white
1977	1:20,000 aerial photography, black and white
1982	1:40,000 aerial photography, colour

Figure 2.10 Example of a 'metadata' map showing information on the currency of the source material used to compile the original map.

For many GIS users the inclusion of metadata has not been seen as particularly important in the creation of spatial databases. However, as more spatial datasets are created and versions of these datasets proliferate the need for good quality metadata increases. It is easy to conceive of the situation where multiple copies of the same information, each slightly different, are being used within the same organisation. Lillywhite (1991) summarised the metadata needs of the GIS user for each stage of a project (Table 2.2).

Map users would expect to find on the printed map information about the currency of the map with publication and print dates, as well as information about the history of that particular edition such as the compilation date for the topographic base, the last time the road information was revised and so on. This information may even be formalised in some way. The military maps produced by the North Atlantic Treaty Organisation (NATO) countries are carefully co-ordinated so that an edition produced by the US Defense Mapping Agency and later updated by the UK's Military Survey will have the edition number changed from 1-DMATC to 2-GSGS. The Ordnance

STAGE - METADATA TASK	METADATA NEEDED
Define scope of project Identify possible datasets	Catalogues and directories of data holdings, prices summary

Define project tasks
    Check possible datasets
    for suitability for tasks,
    validity (within a dataset)
    and consistency (between
    datasets)
	Detailed textual description of data
	in user terms (i.e. geographer,
	epidemiologist, surveyor, civil
	engineer etc.)
	
	Information quality and
	completeness, history and plans
	for updates and corrections.
	Information on assumptions,
	granularity, original scale.

Organise data processing
    Understand/define syntax
    semantics of file structures,
    define data transformations.
    Specify correction for
    distortions.
	Description in formal programming
	terms (entity and attribute analysis).
	Formal specification language (e.g.
	VDM, see Ruggles *et al.*, 1990).

Arrange data transfer
    Negotiate schedules, prices
    and other contractual details.
	Media, logistic constraints, data
	transfer procedures, price details,
	terms and conditions, caveats.

Table 2.2 The metadata needs for each stage of a GIS project
(Lillywhite, 1991)

Survey use a coding system that uses letters to indicate new editions, bars under the letters to show major changes that are not significant enough to imply a new edition, and asterisks adjacent to the letters to show minor amendments.

Very few maps produced from digital mapping or GIS systems contain information about the currency and quality of the data. The inclusion of metadata information is not an aspect in the design of many systems which has been given much consideration. However, to understand fully the limitations on any type of spatial query or analysis it is necessary to appreciate the underlying data in terms of its accuracy, currency and context.

The successful utilisation of spatial data for decision making requires the inclusion of metadata in both the graphic and the non-graphic databases. The metadata must then be easy to obtain, especially if they are not automatically presented to the user as part of the information display, either on the map or with associated tabular data.

Spatial information users must be made more aware of the limitations of taking GIS information at face value. By not addressing the overall quality of the information, the consequences of inaccuracies in any one dataset on the outcomes of the decision- making process will not be taken into account. Issues of data quality, principally in regard to spatial location, were addressed in more detail in Goodchild and Gopal (1989). There is a need to establish standards on the ways of describing the existence, content, format and location of data to effect better user awareness of spatial data within large integrated information systems Burnhill (1991).

References

Burnhill, P. (1991) Metadata cataloguing standards. *Proceedings of symposium on 'Approaches to the Handling of spatial metadata'.* Ed. Medyckyj-Scott, D., Association for Geographic Information, London, pp. 10-16.

Cassettari, S., Fagg, A, and Visvalingham, M. (1992) *Cartography and GIS.* Education and Training Research Publication Number 5, Association for Geographic Information, London.

Cornelius, S. (1991) Data auditing. *Proceedings of Symposium on 'Approaches to the Handling of spatial Metadata.'* Ed. Medyckyj-Scott, D., Association for Geographic Information, London, pp. 17-22.

Dale, P. F. and McLaughlin, J. D. (1988) *Land Information Management: An Introduction with Special Reference to Cadastral Problems in Third World Countries.* Oxford University Press,

Oxford.

Dangermond, J. (1979) A case study of the Zulia regional planning study, describing the work completed. *Harvard Library of Computer Graphics*, Vol 3, pp. 35-62.

Fisher, P. F. (1991) Spatial data sources and data problems. In:*Geographical Information Systems Vol 1: Principles* , Eds. Maguire, D. J., Goodchild, M. F. and Rhind, D. W., Longman, London, pp. 175-189.

Flowerdew, R. (1991) Spatial data integration. In *Geographical Information Systems Vol 1: Principles*. Eds. Maguire, D. J., Goodchild, M. F. and Rhind, D. W., Longman, London pp. 375-387.

Flowerdew, R. and Green, M. (1989) Statistical methods for inference between incompatible zonal systems. In: *Accuracy of Spatial Databases*. Eds. Goodchild, M. F. and Gopal, S., Taylor and Francis, London, pp. 239-47.

Goodchild, M. F. and Gopal, S. (eds) (1989) *Accuracy of Spatial Databases*. Taylor and Francis, London.

Lillywhite, J. (1991) Introduction: The problem. *Proceedings of Symposium on 'Approaches to the Handling of spatial Metadata'* Ed. Medyckyj-Scott, D., Association for Geographic Information, London, pp. 6-9.

Monmonier, M. S. (1977) Maps, distortion and meaning. *Association of American Geographers Resource Paper 75-4,* AAG, Washington DC.

Openshaw, S. (1983) Multivariate analysis of census data: the classification of areas. In: *A Census User's Handbook*. Ed. Rhind, D. W., Methuen, London, pp. 243-64.

Openshaw, S. (1984) The modifiable areal unit problem. *CATMOG 38.* GeoBooks, Norwich.

Openshaw, S. (1991) Developing appropriate spatial analysis methods for GIS. In: *Geographical Information Systems Vol 1: Principles*. Eds. Maguire, D. J., Goodchild, M. F. and Rhind, D. W., Longman, London, pp. 389-402.

Raper, J., Rhind, D. and Shepard, J. (1992) *Postcodes. The New Geography*. Longman, London.

Rhind, D. W. and Clark, P. (1988) Cartographic data inputs to global databases. In: *Building Databases for Global Science*. Ed. Mounsey H. M., Taylor and Francis, London, pp. 79-104.

Ruggles, C., Medyckyj-Scott, D., Newman, I and Walker, D. (1990) Data structures for identifying and locating relevant spatial information. *Proceedings European Conference on Geographical Information Systems*, Amsterdam, pp. 958-67.

Visvalingham, M. (1990) Trends and concerns in digital cartography. *Computer-Aided Design,* Vol 22, No 3, pp. 115-130.

3

Geo-information databases

3.1 Choosing the right data structure

One of the limitations of using GIS for effective decision making is the structure of geo-databases which have in the past been designed principally to generate cartographic products. Given that the basis for GIS is the need to understand the spatial context of features and events, the criteria for the storage and management of spatial information should be more than just the representation of that data in a cartographic form.

The focus of GIS in recent years has been on the capture of existing map information and its subsequent display, often in a form very much like that of the paper product. This has constrained the development of spatial data models that would be more suited to spatial information handling and analysis in an integrated decision support framework.

This is not to suggest that the map as a form of communicating spatial information is not a very important component in GIS. The problem in the development of spatial decision support systems is the need to define suitable data structures and data models for developing interactive information manipulation as part of a broad and comprehensive approach to problem solving.

3.2 Data structures and data models

In designing information handling strategies it is necessary to distinguish between the data structure and the data model. These have also been referred to as low level and high level data structures respectively (Egenhofer and Herring, 1991). The data structure is the method adopted for organising the spatial data, such that the following are addressed:

- All the appropriate geometric locations and relationships are stored.
- The data are organised so that the various types of query can be

answered.
- Results from data manipulations are consistent and repeatable.
- The computer processes are efficient.

The data model on the other hand is the definition and formalisation of the semantics for a particular spatial concept (Egenhofer and Herring, 1991). The use of geographic information can embrace a number of spatial concepts. For example the location of furniture in a room and a person's ability to move around the room without hitting the furniture is based on the knowledge that each object has a known size and location within the geographic space that is the room. This is based on Euclidean geometry in which continuous space consists of an infinite number of points. The concepts of space and their measurement are discussed more fully in Gatrell (1991). An alternative spatial concept is that adopted when selecting the route along which to travel involving the use of spatial networks.

In very simple terms, the difference between the data model and the data structure is that the model lays down a set of rules by which we deal with geographic space, while the structure details how these rules are implemented in a computer system. A growing literature exists which considers these issues in more detail (see Mark *et al.* 1989). The question for the information manager is the required level of understanding about the the data models and data structures necessary for effective use of a GIS.

3.3 Spatial data models

Various data models have been developed to represent geographical space. The vector and raster models form the simplest level. The vector is a representation of the spatial entities which make up the geographic space. Most commonly the model is enhanced by utilising the spatial concept of topology, which describes an object's relative rather than actual location. Figure 3.1 shows three spatial representations in which the geometry of the objects is all different but the topological relationships in A and B are the same.

The raster model divides the space into equal sized cells and is implemented by the use of a number of data structures such as quadtrees and run-length codes, referred to in the next section. Some of the issues relating to topological models and the appropriateness of vector versus raster data are to be found in Green *et al.* (1985), Smith *et al.* (1987) and Laurini and Thompson (1992).

The advances in GIS during the 1970s and 1980s focused on the development of data models for handling cartographic representations of

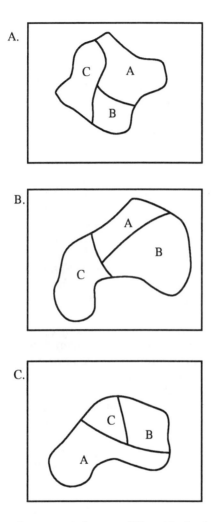

Geometric representations are different in the three graphics but
the topological relationships are the same in A and B.

Figure 3.1 Topological representations of geographic objects.

spatial phenomena and the creation of linkages between these models and database management systems. The best-known example is that of ARC/INFO, which comprises an integrated solution linking a cartographic package, ARC, which manages the X,Y co-ordinate data in the form of geographic coverages equivalent in concept to the single map sheet, and a relational database management system, INFO, which stores values associated with the cartographic components (Aronson and Morehouse, 1983). One of the preconditions for the design of the ARC/INFO model was that each element should utilise existing data handling models, thereby allowing technical developments to be incorporated into the system (Morehouse, 1985).

The result of this approach to the creation of spatial information databases is to distinguish between the geometric representation of the spatial entity and the descriptive or attribute data of various types that may be associated with this. As a consequence many GIS data models have adopted a vector approach and represent the world as a series of basic cartographic units, such as a point or line, which have a location in space defined by some form of co-ordinate geometry and to which may be added various forms of identifiers and descriptors. The topological model used by ARC/INFO is an example. Figure 3.2 shows the basic data structure of the cartographic entities and how their topology is used to create geographic features. This vector data model is usually referred to as the entity-relationship model.

An alternative data model is that of the object-oriented GIS. Various approaches to the application of object-oriented design principles to GIS have been described, for example Chance et al. (1990), Aybet (1992) and Worboys et al. (1990). Newell (1992) notes that 'there does not seem to be a precise definition of an object-oriented system' but that there are degrees of 'object-orientedness' depending on how many of the concepts of object-orientation are supported. In essence, however, the object-oriented model is unlike the entity-relationship model. Instead it is based on modelling objects by their attributes and their relationships to other objects. The benefits of this approach are summarised by Newell (1992) as:

- One object may possess multiple alternative geometries, which may be useful for generalisation.
- One geometry, such as a node, link or polygon may be shared by any number of objects.
- Explicit topology for objects from many themes or classes can be modelled, so one is not limited to single theme coverages.
- It is easier to provide generic support for a much richer data model than is available in the geometry-centred data model.

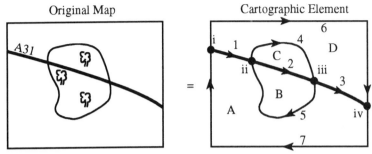

Original Map   Cartographic Element

Direction arrows used to determine start and end points and thus areas to left and right.

Segement	Start Point	End Point	Area to Left	Area to Right
1	i	ii	D	A
2	ii	iii	C	B
3	iii	iv	D	A
4	ii	iii	D	C
5	iii	ii	A	B
6	i	iv	X	D
7	iv	i	X	A

1. To define the object area 'B' select all the segments , with 'B' in the columns 'Area to left' or 'Area to right'.

2. To define the object 'woodland' which is areas 'B' and 'C' select all segments with a 'B' or a 'C' in the columns 'Area to left' or 'Area to right' but not both.

Figure 3.2 Use of a topological data structure for assembling cartographic objects in a vector data model.

3.4 Spatial data structures

It is not the purpose of this book to discuss in detail the various spatial data structures used in GIS. There is however a need to recognise that the choice of spatial data structure can have a profound effect on the decision-making capabilities of an organisation using that particular GIS tool. Even for the most simple data manipulation and display, different structures offer a range of advantages and disadvantages. As the requirements increase for more sophisticated query and analysis functionality, so the need increases to clearly identify the most appropriate method of data handling.

It is possible to convert between one data structure and another without changing the spatial data model. But in practice the data structures for both the spatial entities and their attribute data are related to the way in which the data are to be used and any time and resource constraints imposed on the collection process. As a result, a wide variety of data structures proliferate and an extensive literature exists on the subject. Knapp and Rider (1979) described three polygon data structures, namely the dual independent map encoding (DIME) file structure, the whole polygon structure and the arc and node structure. Triangular structures have been used to represent three dimensional surfaces (Males, 1977 and Gasson, 1983) and a variety of raster data structures exist which use a regular, irregular or hierarchical form (Samet, 1983; Samet 1984). A GIS based on a quadtree cell addressing system was described by Palimaka et al. (1986), and developments of the quadtree, such as bin-trees (B-trees), B+-trees and k-d-trees were considered by Kleiner and Brassel (1986). A detailed discussion of hierarchical data structures can be found in texts by Samet (1990a and 1990b). The use of the quadtree data structure for vector data was discussed by Ibbs and Stevens (1988).

As an example of the importance of appropriate data structures Figure 3.3 shows a simple map coded using run-length codes and chain codes. Various measures may be important in assessing the validity of adopting a particular structure, depending on the stated objectives. In Figure 3.3 the computer storage requirements may be simply assessed by counting the number of characters required to code the map. The type and shape of features is important in determining the storage needs.

Once the information system is in place the data structure may be of little consequence to the users as long as they are able to achieve the desired manipulation and analysis. The most successful data structures are those that offer the greatest flexibility in manipulation, analysis, retrieval and display, while at the same time remaining transparent to the users. An information system which requires a good understanding of the data structure to operate successfully limits its wider usage as an

1. Run length codes

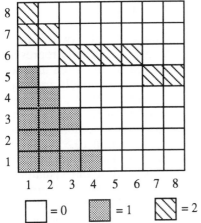

Code:
Line number,value,number of cells

1,2,4,0,4
2,2,2,0,6
3,2,3,0,5
4,2,2,0,6
5,2,1,0,4,1,3
6,0,2,1,4,0,2
7,1,2,0,6
8,1,1,0,7

= 0   = 1   = 2

2. Chain codes

Code:
Start point, direction,number of cells

Area I   1,8,1,7.2,3,
3,2,0,1,3,4,
0,1,3,1,0,1

Area II   0,8,1,1,2,1,1,1,
2,1,1,4,2,1,1,2,
2,1,3,3,0,1,3,3,
0,1,3,2,0,2

Area III   8,8,3,4,0,1,3,2,
0,1,1,1,0,1,3,1,0,1,
3,1,0,1,3,1,0,1,1,2,
2,1,1,3,2,1,1,3,2,4,

Area IV   0,0,0,5,1,1,2,1,1,1,2,1,1,1,1,
2,1,3,1,2,1,1,2,2,2,1,3,4

= 0   = 1   = 2

Convention used : North = 0; East =1; South = 2; West = 3

◄—●   Start point and direction of coding

Figure 3.3 Examples of different data structures for a raster data model.

integrated information system.

3.5 Problems of three dimensional data

Most spatial data models in GIS consider only two-dimensional geographic space. The third spatial dimension is usually represented as an attribute, for example contours are stored as linear features with the attribute being the height value. Digital elevation models are based on a geo-referenced regular grid with height values attached to each cell. This is defined as 2.5 dimensional data since it represents a surface and not a true three dimensional object. A common representation of this type of data is in the form of perspective views (see section 10.8).

In order to create a true three-dimensional data model the data must be manipulated and displayed as a solid structure and visualised in its true perspective. The key analytical functions are those of volume, object intersection and surface calculations. In addition, many of the functions found in two dimensional GIS are also important, such as scaling, transformations, topological relationships and the and/or selection of three dimensional elements. The major limitation on the development of three dimensional spatial models has been the ability of computer hardware to handle the volumes of data and processing requirements. This is gradually changing, leading to the development of applications in geology, environmental monitoring, civil engineering and landscape architecture (see Kelk, 1991; Raper and Kelk, 1991).

The importance of establishing spatial models for three dimensional data will become increasingly important with the development of more sophisticated visualisation techniques. The issue of visualisation and its critical importance for improving information management and decision making is discussed more fully in Chapter 10.

3.6 Temporal data model

A further aspect to the establishment of more widely applicable spatial data models is the integration of the spatial and temporal components. The discussion on maps and other types of spatial data highlighted the importance of temporal change. But there are very few GIS software solutions that incorporate time as anything more than an attribute to spatial entities.

The spatial models need to address changes in both the spatial entities and their attributes through time. Figure 3.4 shows how a river may have a defined spatial location which represents its position at a particular point in time. During the natural course of events the river's

1970   1980

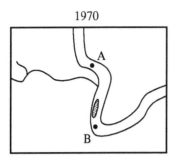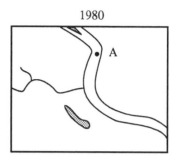

A,B = sampling stations

Station A   Station B

Date	Water Flow
.	.
.	.
.	.
1/6/70	28.3 m³/s
2/6/70	27.9
3/6/70	27.8
.	.
.	.
.	.
1/6/80	26.4
2/6/80	27.5
3/6/80	26.9
.	.
.	.
.	.

Date	Water Flow
.	.
.	.
.	.
1/6/70	29.1 m³/s
2/6/70	28.6
3/6/70	28.4
.	.
.	.
.	.

Figure 3.4 Temporal aspects to position of spatial objects and their attributes.

position will change, thus the 'same' spatial object will have a new spatial location. This object may also have attribute data such as water sampling stations which will generate time series data such as river flow or water quality.

Langran (1992a, 1992b) discusses the need for a conceptual model for cartographic time. In temporal terms the cartographic map is a 'state' defined in terms of general systems theory 'which considers the history of a system to be a series of states punctuated by "events" that transform one state to the next.' (Langran, 1992a). The cartographic state is a representation of spatial objects which is transformed to another state by an event. Each version of an object is transformed to another version of that object by a 'mutation'. Thus the map is a snapshot in time of a number of evolving geographic objects (Figure 3.5).

Depending on the nature of the processes the interaction between objects as they change may be viewed as instantaneous and defined as a point on the temporal scale or be a gradual process in which distinct elements in the version/mutation/version model are difficult to distinguish. An example of the latter might be the mapping of slopes or the changes in a coastal spit through time. This presents us with the problem of defining what constitutes a mutation: is it the deposition of a

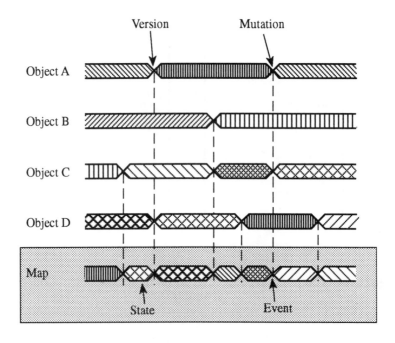

Figure 3.5 The relationships of objects to maps in a model of cartographic time (Langran, 1992a).

single grain of sand or the growth in area of the spit by a particular amount? The model therefore has to define the 'magnitude' of a mutation event for it to be considered significant enough to effect a change in version. This will alter depending on the type of process. Table 3.1 is a summary of the components of cartographic time defined in the model discussed by Langran (1992a). It can be seen that for some GIS applications a more extended model is required where there is some level of interaction between cartographic time and cartographic space.

Very few spatial models incorporate the temporal aspects and yet to fully understand the consequences of implementing specific information strategies, time is a key component. The complexes of spatiotemporal data are such that traditional entity-relationship models may not be adequate. There have been some attempts to develop object-oriented models for spatiotemporal data as an alternative, for example Worboys (1992).

3.7 Database management systems

The development of suitable models and data structures for spatial data in GIS has been paralleled by the need to develop suitable database management systems (DBMS) for the creation of spatially referenced databases, or geo-databases. Many of the functions that exist in databases designed for non-spatial data are equally applicable to GIS applications but there are also a number of limitations with existing DBMS with respect to the successful manipulation of spatial data

---

- Time is a fourth cartographic dimension
- Cartographic time and cartographic space do not interact
- Temporal boundaries are sharply drawn
- The language of maps extends to temporal constructs
- Spatial objects have temporal corollaries
- Cartographic time's topology is that of many parallel lines, one per object
- Cartographic time has three facets: world time, database time, and display time
- Objects maintain their identity despite change over time
- Objects can change in different ways through time
- Temporal objects can be treated irrespective of scale

Table 3.1 The components of cartographic time (Langran, 1992a).

(Healey, 1991).

Databases are a structured approach to the storage of a number of data files. Relationships between the different files and individual data elements contained in them may be established for the purposes of improving manipulation and retrieval of information related to specific problems. A most important basis for the development of DBMS is their ability to service the needs of a variety of users. To achieve this the DBMS has to be flexible both in its structure and the ways in which data are retrieved. As a result the DBMS becomes a tool for more efficient management of an organisation's data resource. For a detailed introduction to database management systems the reader is referred to Date (1990) and McFadden and Hoffer (1991).

The most common approach to the use of DBMS for spatial data is to adopt a hybrid approach in which the geographic data are stored in their own data structure as part of a GIS software solution, with the attribute data in a standard DBMS which is linked in some formalised way to the geographic data structure (Batty, 1991). This has some advantages in that the data structure for the geographic data can be optimised for specialised data handling functions such as cartographic display or polygon overlay. It also means that existing databases can be used to store attribute data.

The alternative approaches to this 'hybrid' form of GIS database are to create a distributed database or to store all aspects of the spatial data in a single DBMS. Distributed databases are potentially the most attractive, since GIS is an integrating information management tool which can link data from apparently disparate sources. A distributed database is the key to integrated information, potentially accessing data from a number of computer systems, both within a organisation and across organisations, in a way that is transparent for the user. However to achieve this while maintaining the integrity of the data and the manipulation processes is an extremely complex task. Batty (1992) considered that the current consensus is to use a single DBMS for all GIS data, which simplifies access. This can help solve problems of various users wishing to access spatial datasets which are in various stages of revision or change, problems of ownership for derived datasets and the security of the basic data.

3.8 Physical and logical database design

To implement an information management system successfully while utilising one or more than one DBMS for the storage and retrieval of data requires careful consideration of the design of the individual databases and how these are to be related in order to achieve the necessary data access and manipulation for the defined user group.

There are two aspects that need to be considered. The first is the physical design, or the actual computer location of a dataset. It may be that different spatial data attributes are collected and input into a database by different individuals, and it may be desirable to create separate data files located on different computer storage devices in order to balance the input and maintenance procedures or to reduce the effects of a system failure.

The second is that of logical design. This is required to establish clear relationships between the specific data elements or datasets in order to define the conceptual information model. Failure to establish the correct conceptual model invariably leads to an inefficient database structure, unnecessary data redundancy in storage and a poor match between user's requirements for data access and manipulation (Healey, 1991).

There are various techniques that have been used for establishing the most appropriate model, but that of the entity-relationship model approach (Chen, 1976) has been most widely accepted. The example in Figure 3.6 shows that there is a relationship between the basic entity and various subsets of that entity, and that the relationships may be defined in terms of one-to-many or many-to-many. In this example the basic entity is the park, but could be any generic entity set such as towns, postcode areas or land-use units. Each of the entities in the set have a number of defined characteristics or attributes. The entity set may be related through specific relationship sets, so that accessible park sites will have more than one vegetation type and will be accessible by more than one trail. It is also possible to define a one-to-one relationship, such as the towns along a railway which have one, and only one, railway station.

3.9 DBMS structures

A large number of DBMS software packages are available which can be organised to store the data defined in the conceptual model. These can be broadly categorised in the following way:

- *Inverted list systems.* This is the simplest form and utilises tables or lists of data in columns and rows. The ordering of the rows and columns affects data access.
- *Hierarchical systems.* This is the world's most widely used system, based on the IBM IMS (Information Management System), first released in 1968. One entity set forms the root of the hierarchical tree with a series of pointers to the next level

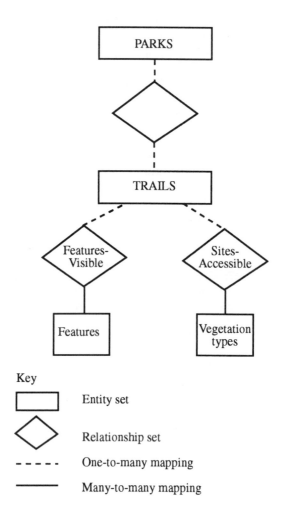

Figure 3.6 An example entity-relationship model for a national parks database (Healey, 1991).

down to represent the relationships between types of entity sets. This does not handle the many-to-many relationship well.

- *Network systems.* Also referred to as CODASYL databases (from the Database Task Group of the Conference on Data

System Languages), the difference between network and hierarchical models is that in the network systems one entity set may be linked to another entity set, including more than one at a higher level.

- *Relational systems.* The concepts of relational systems were first proposed by Codd (1970). They are characterised by a simple structure with all data in tables of rows and columns. The tables are the relations between entity sets. This is very similar to the entity-relationship model.

The relational system has become the most commonly adopted for current DBMS and for GIS applications. While it is generally suited to many types of spatial data query it does have some drawbacks. The main ones is the limitation of the standard query language (SQL) for use in geographical data types (see Section 3.10).

The alternative to these structures are those based on the hybrid, using a DBMS for the attribute data and the GIS for the co-ordinate geometry. This is based on the view that data storage structures cannot be optimised for both the locational information and the attribute data (Morehouse, 1985). The key to this approach is the use of unique identifiers between the co-ordinate files and the attributes in the database which allow attributes to be tied to spatial elements. There are a number of GIS solutions based on this hybrid approach. Most use relational databases, although there are examples of systems that use other database types such as the Intergraph IGDS/DMRS, which uses network DBMS. The main differences in these systems is the structure of the cartographic or co-ordinate file structure, depending on whether it is a CAD-based system or a vector-topological system (see Healey, 1991).

A further model is the integrated approach or spatial database management system approach. Here the GIS sits directly on top of the database and is used to query both the co-ordinate data and the attribute data in the database. Most of these types of geo-database use the vector-topological data model (Morehouse, 1989). The attributes and the co-ordinate data are stored in the same tables as the map features or separate tables linked by relational linkages. This provides for a more convenient way of retrieving data, particularly when little or no analysis is to be performed, such as for display purposes. This type of model has been utilised for large scale cartographic databases by both the US Geological survey (Starr and Anderson, 1991) and the UK Ordnance Survey (Sowton, 1991). The development of the integrated approach will depend on the continuing developments in distributed computing, advances in computer networks and the development of GIS functionality for a multi-user environment to match that being developed for other DBMS.

3.10 Structured query language (SQL)

A query language is the tool a user needs to extract data from a database and present the result in a useful format (Frank and Mark, 1991). The structured query language (SQL) is used extensively in many database applications and has become a standard for relational DBMS. It was designed as a high level interface for the manipulation of tables and has been successfully used in standard applications such as accounting. Its use for geographical data management has some limitations, the most significant of which is the inability to define geographical data types and operators (Batty, 1992). For example it is not possible to select a subset of data based on geographical area.

Various extensions to SQL have been proposed, including SQL2 and SQL3, but at present there exists no satisfactory SQL-based interactive GIS query language. Egenhofer (1992) identifies the two main deficiencies of spatial SQL as 1) the severe difficulties in incorporating spatial concepts such as graphical display and specification, and 2) the lack of power in the relational framework which will not support qualitative answers, knowledge queries and metadata queries. All of these are important in handling spatial information, in particular the role of metadata that was discussed in Chapter 2.

Some would argue that the developments to SQL do not represent the solution to the problems presented with spatial data. Proposed amendments suggest that SQL will be turned into a 'complete' programming language leading to an increase in the complexity of SQL (Egenhofer, 1992). The alternative view is that the proposals for user-defined data types and operators included in the draft SQL3 standard will make the handling of geographic data using SQL significantly easier (Dowers, 1991). The argument for improving the existing SQL standard is based on its current widespread usage rather than choosing something completely new. It is likely that SQL3 or a variant will contain adequate query capabilities for geo-information for current GIS applications, but with developments to the spatial models a new GIS query language will be required.

3.11 Object-oriented databases

In the discussion on spatial data, we identified the need to address data not merely in terms of cartographic elements but as spatial entities. This is reflected in the development of object-oriented data structures and techniques, such as the development of POSTGRES, an object-oriented successor to the relational DBMS INGRES, the development of the Intergraph TIGRIS system which utilises an object-oriented approach to programming or the Smallworld GIS described by Newell (1992).

Newell (1990), in an overview of the concepts, summarises the advantages to the developer as being the ability to alter parts of the database system without causing unfortunate side effects, and the use of the concept of inheritance means that generic code can be developed over and over again. The result to the user should be a much more powerful and extendable system.

There a number of limitations to the implementation of object-oriented concepts for spatial information management. Date (1990) identifies such issues as the problems of creating a general query language for an object-oriented database and the difficulties of creating a distributed form of the database. The object-oriented approach also has some practical constraints such as the lack of established skills, the limited acceptance within the GIS industry and the problem of integrating with existing spatial data. Another significant problem is the lack of established standards. The benefits of establishing the widely accepted standards for the GIS community is extremely important in establishing the wider goals of information management (see Chapter 7).

References

Aronson, P. and Morehouse, S. (1983) The ARC/INFO Map Library: A design for a digital geographic database. *Proceedings 6th International Symposium on Automated Cartography*, Ottowa, pp. 372-82.

Aybet, J. (1992) Object-oriented GIS: What does it mean to GIS users? *Proceedings European Geographical Information Systems Conference*. Munich, pp. 1279-87.

Batty, P. M. (1991) GIS databases - which way forward?. *Proceedings AM/FM European Conference VII*, Montreux.

Batty, P. M. (1992) GIS databases - A distributed future. *Mapping Awareness* , Vol 6, No 5, pp. 34-7.

Burroughs, P. A. (1986) *Principles of Geographical Information Systems for Land Resources Assessments.* Clarendon Press, Oxford.

Chance, A., Newell, R. G. and Theriault, D. G. (1990) An object-oriented GIS - Issues and solutions. Proceedings European Geographical Information Systems Conference, Amsterdam, pp. 179-88.

Chen, P. (1976) The Entity-Relationship Model - towards a unified view of data. *Association for Computing Machinery transactions on*

Database Systems, Vol 1, No 1, pp. 9-36.

Codd, E. F. (1970) A relational model of data for large shared data banks. *Communications of the Association for Computing Machinery* , Vol 13, No 6, pp. 377-87.

Date, C. J. (1990) *An Introduction to Database Systems,* Volume 1. Addison Wesley, New York, Fifth Edition.

Dowers, S. (1991) SQL - The way forward. *Proceedings of AGI 91 Conference,* Birmingham, pp. 3.13.1-4.

Egenhofer, M. J. (1992) Why not SQL? *International Journal of Geographical Information Systems ,* Vol 6, No 2, pp. 71-85.

Egenhofer, M. J. and Herring, J. R. (1991) High-level spatial data structures for GIS. In: *Geographical Information Systems Volume 1: Principles.* Eds. Maguire, D. J., Goodchild, M. F. and Rhind, D. W., Longman, London, pp. 227-37.

Frank, A. U. and Mark, D. M. (1991) Language issues for GIS In: *Geographical Information Systems Volume 1: Principles.* Eds. Maguire, D. J., Goodchild, M. F. and Rhind, D. W., Longman, London, pp. 147-63.

Gasson, P. (1983) *Geometry of Spatial Forms.* Ellis Horwood, Chichester.

Gatrell, A. C. (1991) Concepts of space and geographical data. In: *Geographical Information Systems Volume 1: Principles.* Eds. Maguire, D. J., Goodchild, M. F. and Rhind, D. W., Longman, London, pp. 119-34.

Green, N. P., Finch, S. and Wiggins, J. (1985) The 'state of the art' in Geographical Information Systems. *Area ,* Vol 17, No 4, pp. 295-301.

Healey, R. G. (1991) Database management systems In: *Geographical Information Systems Volume 1: Principles.* Eds. Maguire, D. J., Goodchild, M. F. and Rhind, D. W., Longman, London, pp. 119-34.

Ibbs, T. J. and Stevens, A. (1988) Quadtree storage of vector data. *International Journal of Geographical Information Systems ,* Vol 2, No 1, pp. 43-56.

Kelk, B (1991) 3-D GIS for the geosciences. *Computers and Geosciences* 17

Kleiner, A. and Brassel, K. E. (1986) Hierarchical grid structures for static geographic databases. *Proceedings Auto Carto London,* Vol 1, pp. 485-96.

Knapp, E. M. and Rider, D. (1979) Automated geographic information systems and Landsat data: a survey. *Harvard Library of Computer Graphics,* Vol 4, pp. 57-68.

Langran, G. (1992a) *Time in Geographic Information Systems.* Taylor and Francis, London.

Langran, G. (1992b) States, events and evidence: The principle entities of a temporal GIS. *Proceedings GIS/LIS*, California, pp. 416-9.

Laurini, R. and Thompson, D. (1992) *Fundamentals of Spatial Information Systems*. Academic Press, London.

McFadden, F. R. and Hoffer, J. A. (1991) *Database management*, Edition 3. Benjamin/Cummings, Redwood city.

Males, R. M. (1977) ADAPT - A spatial data structure for use with planning and design models. *Proceedings Advanced Symposium on Topological Data Structures for Geographic Information Systems*, Vol 2, pp. 1-35.

Mark, D. M., Frank, A. U., Egenhofer, M. J., Freundschuh, S., McGranaghan, M. and White, R. M. (1989) Languages of spatial relations: report on the meeting for NCGIA Research Initiative 2. *Technical Report 89-2*, National Center for Geographic Information and analysis, Santa Barbara.

Morehouse, S. (1985) ARC/INFO: A geo-relational model for spatial information. *Proceedings Auto Carto 8*, ASPRS, Falls Church, Virginia, pp. 388-97.

Morehouse, S. (1989) The architecture of ARC/INFO. *Proceedings of AUTOCARTO 9*, ASPRS, Falls Church, Virginia, pp. 266-77.

Newell, R. G. (1990) Who cares about object-orientation? *Mapping Awareness*, Vol. 4, No. 4, pp. 10-11.

Newell, R. G. (1992) Practical experiences of using object-orientation to implement a GIS. *Proceedings GIS/LIS Conference*, San Jose, pp. 624-9.

Palimaka, J., Halustchak, O. and Walker, W. (1986) Integration of a spatial and relational database within a geographic information system. *Proceedings ACSM/ASP Annual Convention Technical Papers*, Washington, Vol 3, pp. 131-40.

Raper, J. F. and Kelk, B. (1991) Three-dimensional GIS. In: *Geographical Information Systems Volume 1: Principles*. Eds. Maguire, D. J., Goodchild, M. F. and Rhind, D. W., Longman, London, pp. 299-317.

Samet, H (1983) Hierarchical data structures for representing geographical information. *Proceedings US/Australian Workshop 'The Design and Implementation of computer-Based Geographic Information Systems*, Hawaii, pp. 36-50.

Samet, H. (1984) The quadtree and related hierarchical data structures. *ACSM Computing Surveys*, Vol 16, pp. 187-260.

Samet, H. (1990a) *The Design and Analysis of Spatial Data Structures*. Addison-Wesley, Reading, Massachussetts.

Samet, H. (1990b) *Applications of Spatial Data Structures: Computer Graphics, Image Processing and GIS*. Addison-Wesley, Reading, Massachussetts.

Smith, T. R., Menon, S., Star, J. L. and Estes, J. (1987) Requirements and principles for the implementation and construction of large-scale geographic information systems. *International Journal Geographical Information Systems,*Vol 1, No 1, pp. 13-31.

Sowton, M. (1991) Development of GIS-related activities at the Ordnance Survey. In: *Geographical Information Systems Volume 2: Applications*. Eds. Maguire, D. J., Goodchild, M. F. and Rhind, D. W., Longman, London, pp. 23-38.

Starr, L. E. and Anderson, K. E. (1991) A USGS perspective on GIS. In: *Geographical Information Systems Volume 2: Applications*. Eds. Maguire, D. J., Goodchild, M. F. and Rhind, D. W., Longman, London, pp. 11-22.

Worboys, M. F. (1992) Object-oriented models for spatiotemporal information. *Proceedings GIS/LIS,* San Jose, pp. 825-34.

Worboys, M. F., Hearnshaw, H. M. and Maguire, D. J. (1990) Object-oriented data models for spatial databases. *International Journal of Geographical Information Systems,* Vol 4, No 4, pp. 369-83.

4

Integrated approaches to GIS

4.1 Benefits of information integration

The key to successful information management in a spatial context is the integration of spatial and non-spatial data. Maguire (1991) states that spatial information systems are best described as an integrated collection of hardware, software, data and people which operate in an institutional context. The broad systems context of spatial information systems has already been addressed, together with the data issues and various spatial data models. In this chapter the development of strategies for achieving a fully functional and integrated spatial information management system will be reviewed.

There are a number of benefits to be gained from adopting a strategy of information integration, and developing the necessary operational systems and technological solutions. The benefits for the individual or project team were summarised by Shephard (1991):

- A broader range of operations can be performed on integrated information than on disparate sets of data.
- By linking data sets together, spatial consistency is imposed on them. This adds value to existing data, making them both a more effective and a more marketable commodity.
- Through the integration of data which were previously the domain of the individual disciplinary specialists, an interdisciplinary perspective to geographical problem solving is encouraged.
- Users benefit from the perception that they have access to a seamless information environment, uncomplicated by the need to consider differences in data sources, information types, storage devices, computer platforms etc.

From the individual's point of view, these benefits lead to greater interest and motivation in the tasks they are to perform and can result in increased efficiency and a better quality of problem solving. It is, however, important that individuals or project teams have clearly identified aims and objectives, with well-defined strategies for achieving these. If they do not, there is the possibility that greater amounts of information simply extend the time taken to achieve results that are adequately arrived at with less information and a simpler approach.

A further set of benefits can accrue for the organisation (Bracken and Higgs, 1990). The basis of the benefits is greater awareness of the tasks being performed and the information being used for those tasks. This leads to a reduction in data collection costs and prevents duplication of capture and manipulation processes. The inclusion of additional processes into the task of one individual, which may not be of any direct benefit to the successful completion of that individual's task, may be of considerable benefit to other parts of the organisation both in efficiency and cost terms.

A second benefit is the possibility for undertaking tasks using the collective information resource which were previously not possible because of the fragmented nature of information management. As an example, the conservation section of the local authority, which is typically part of the planning department, will have better access to the programmes and plans of the department responsible for parks and open spaces. This will give the conservationists an opportunity to plan a strategy that fits within the objectives of the parks service.

A further benefit may accrue if organisations are able to make use of each other's information. In commercial terms this can lead to increased opportunity for developing datasets that have a greater market potential. The data collection agencies can thus produce better quality data in the knowledge that investment costs can be recovered.

4.2 Integrated data models

At the lowest level, information integration may take place within a GIS and be concerned with spatial data. The most important level of integration is that between different data models. This has long been an operational problem and only in recent years have there been concerted efforts to create systems that can handle both vector and raster data models.

Some solutions simply adopt display strategies that allow vector data and raster images to be overlaid. This has become possible with improvements in the display technology, both hardware and software, which allow the data to be merged and displayed simultaneously (Antenucci *et al.*, 1991).

More complex approaches facilitate the translation of data from one model to another in a manner that is relatively transparent to the user.

This is necessary for various spatial manipulation and analysis operations which use both vector and raster data. The ways in which vector and raster data are integrated is shown in Figure 4.1 (after Shephard, 1991).

While the technology may exist for effecting vector/raster data integration, careful consideration has to be given to the strategies adopted for achieving this. Many GIS solutions are essentially vector or raster systems to which additional software modules or packages have been added to allow for the handling of the other data model. Some of the existing solutions are quite cumbersome in terms of processing time and the type and variety of inputs required from the user.

In the past an organisation has had to consider which is the most important data model for the majority of information that they handle and for the types of analysis they wish to undertake. They have then selected a GIS which optimises this. Adding functionality to handle the other data model has often not been cost effective for the limited use of those data.

Today many systems allow for the storage and display of both raster and vector data. This simplifies the problem for many users who wish to use both data types. The increased use of image-based data will lead to the development of integrated software solutions (a subject discussed more fully in Chapter 11). The translation between data models is still not a standard feature of many GISs, which limits the manipulation and analysis capabilities.

Where an organisation uses a lot of both vector and raster data and can invest in complex GIS systems that will allow translation between models, the issue becomes one of data management. It is important to ensure that when a vector file is updated the raster translation is also updated, or at least marked as superseded. Translation processes may take some considerable processing time so it may not always be desirable to update both vector and raster versions of a dataset. This requires careful management to ensure issues or currency and quality are addressed consistently across an organisation.

Data formats and the need for widely used standards is also important to the effective integration of data from different data models and from different sources. Many GIS systems now have a wide range of translators to handle a variety of data formats. The need for effective data standards is discussed further in Chapter 7.

4.3 Information networks

The ability to achieve truly integrated information systems depends on establishing the appropriate information network and the hardware

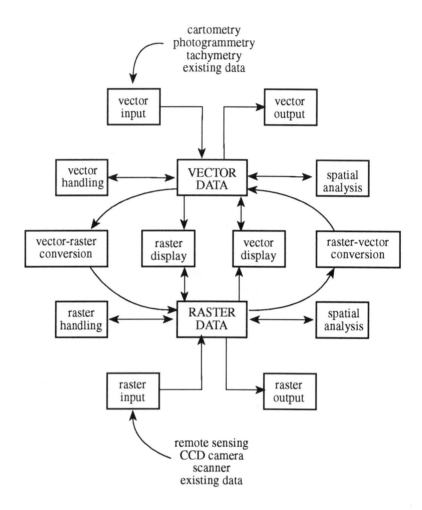

Figure 4.1 Vector and raster data integration (Shephard, 1991).

solutions to operate it. This will depend on the specific tasks carried out by an organisation and the way operational systems have been established to undertake these tasks.

A major problem for many organisations is the development of a configuration for the computer network which directly relates to both the organisational structure and operational structure. Where this is achieved the computer system becomes a complementary asset to information management, but conversely a poor configuration can obstruct efficient information handling. Problems arise in designing a configuration that is flexible enough to meet the needs of an organisation which is constantly changing.

Most computer networks are based on one of four models (Antenucci *et al.*, 1991):

- Simple star network, with a single central node to which all other nodes are connected. This implies a level of centralised control of data and/or applications, and is typically found where the nodes are dumb terminals.
- Tree network, which is a variation on the star network in which there are more than one central node, each of which run several other nodes. The central nodes are connected in some form.
- Ring network, in which all the nodes are located on a ring and can receive and send messages. Information flows in one direction and recognises the target node.
- Linear or BUS network, in which the nodes are located on a 'backbone'. All can send and receive messages and the data can flow in both directions. The most common example of a linear network in GIS configurations is local area networks using the Ethernet standard (Figure 4.2).

There are a number of issues related to the establishment of computer networks which have to be resolved if efficient information management is to take place. These can be broken down into two categories; designing the configuration and managing the resulting system.

When designing the configuration of a computer network the following need to be addressed:

- The physical environment, such as the building or buildings to be serviced by the network, and how far the network has to extend.
- Communication facilities that already exist and can be used, or that have to be integrated.
- The location of existing databases and the responsible department or individual.
- New databases to be created, their size, location and management.
- Applications software and system management requirements.
- Network parameters, such as volumes of data transfer, speed of

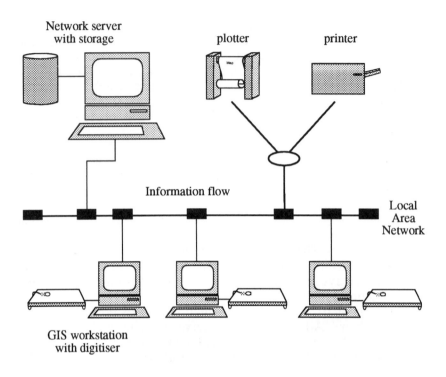

Figure 4.2 Example of a linear or BUS network for a GIS.

response required and numbers of users.
- Security to cover access to the network itself, to particular applications packages which are not for open access and data. Data security includes controlling read/write access to files.
- The need to anticipate future growth in demand and therefore potential upgrade paths for the network.
- Organisational issues, such as structure, perceived aims and objectives and staff requirements.

Managing the network involves a range of tasks:

- Database maintenance, including the control of internal update processes and the update of third party datasets.
- User information services on availability of applications and systems.
- Maintenance of metadata systems, especially currency and

accuracy information.
- Security, including unauthorised access.
- System accounting to identify usage trends and plan for upgrades.

Computer networks can be extremely sophisticated and require high levels of technical knowledge to implement and operate effectively. However, the successful operation of the computing aspects of the network must not be to the detriment of the information aspects. In this context, the user has an important role to play in determining how the network operates.

The development of open systems is also important in this context. The revolution in inter-connectivity between computer systems holds out the promise of reduced costs of initial purchases and upgrades, reduced data capture costs, reduced training requirements and management costs, and a reduction in the risks of being 'tied' to one hardware vendor (Leeson, 1992).

No one system will have all the answers but there is a trend towards developing standards for integration purposes. The X standard is widely available in GIS systems and MOTIF, which uses X, is endorsed by the Open Systems Foundation (OSF) and Unix Systems Laboratories (USL) as an interface standard. There is also the emerging distributed computing environment (DCE) standard which should be important in application development.

4.4 Integration with existing databases

It is widely acknowledged that the full potential of using spatial data is achieved in an integrated framework. Even so, progress towards solutions that may be truly regarded as distributed, integrated and corporate in nature has been slow. This is despite the recognition that 'value added' data can be of enormous benefits to an organisation.

One of the most important aspects of integration which has to be addressed in achieving a distributed system is that of interfacing the GIS to other database management systems (DBMS), in particular relational DBMS. Most GIS products are still based on a 'hybrid' architecture where the spatial information is managed within a proprietary file system and the attribute information is managed within an RDBMS (McLaren, 1990). This limits the degree of integration that can be achieved.

McLaren and Healey (1992), noting that most GIS vendors now 'purport to support the full integration of external RDBMS' with their products, identified a number of key corporate integration requirements and assessed these by reviewing a representative sample of GIS solutions. Their findings are summarised in Table 4.1.

A firm conclusion from these results is that the main tools are available to achieve a reasonable level of integration with available systems, but there are limitations in the areas of data location transparency and data integrity.

In recent years there has been a lack of progress in the creation of systems that allow the GIS to act as a 'spatial data server' and be accessed from external databases. This is being addressed by the development of spatial extensions to the structured query language (SQL) in the form of new standards, SQL2 and SQL3 (Bundock and Raper, 1992).

4.5 Open systems environment

The development of integrated spatial information solutions which help decision makers solve semi-structured problems by allowing the user to choose a solution from a set of workable alternatives were characterised in Chapter 1 as spatial decision support systems (SDSS). The integrating of GIS with other application tools to create an SDSS requires the system designer to make choices in order to optimise flexibility and performance.

To achieve this integration there is a move towards an open systems environment (OSE), which is 'a conceptual framework that describes system components required to provide portable, inter-operable and extendable tools within an open, distributed computing environment' (Davidson, 1992). The OSE model extends the concept of standardising open systems interconnection (OSI) technology, mentioned in section 4.3.

Davidson (1992) describes the OSE reference model as consisting of entities and interfaces, as shown in Figure 4.3. The entities are:

- Application software, which includes data, documentation, training materials and software programs that support a particular activity.
- Application platform, which includes hardware and software components that provide the services used by the applications programs. These include the programming tools that make the platform transparent to the application.
- Platform external environment, which consists of any system components that are external to the application platform or software, such as users and data communication.

Requirement	Review Results
• Interface is SQL based	All systems used SQL
• Variety of DBMSs supported	Only Ingres and Oracle were supported by all systems reviewed
• Interface update facilities, not just read only operations	All interfaces support read only and write operations
• Generic RDBMS integration tools	All systems used generic integration tools
• Product specific integration approaches	No GIS products modify the GIS/DBMS interface to optimise performance
• Multiple, simultaneous RDBMS support	In theory most systems support access to multiple types of RDBMS but there are few operational systems
• External references to data managed within and referenced from the GIS to be included in a data dictionary	None of the GIS products supported a true data dictionary approach
• Full transparency of DBMS integration	GIS products succeeded in hiding the DBMS complexities from the users for most functionality, but required varying levels of customisation
• Distributed DBMS (DDBMS) supported	None of the products reviewed supported DDBMS

Table 4.1 (Part 1) A review of GIS/DBMS integration requirements (from McLaren and Healey, 1992).

- Elegant recovery on system failure

  Fully automated approach to recovery of both GIS and DBMS is not supported, but recovery tools do exist.

- Access from external DBMS to GIS

  Not available 'off-the-shelf' but some specialised applications have been developed

Table 4.1 *contd*

The interfaces between these three entities are:

- Application Program Interface (API) between the application software and application platform. This may include user interface, information interchange, communications and internal system services.
- External Environment Interface (EEI) between the platform and the external environment. This supports transfer of information between users, external databases and other application platforms.

The importance of this approach is the move towards systems which are designed with integration as the kernel concept. The adoption of the OSE Model for spatial information should lead to the development of a generation of spatial information systems which use new architectures to achieve a far higher degree of system integration.

4.6 Integration frameworks

The development of integrated systems requires the identification of specific frameworks which provide the mechanisms that support the system functionality. Davidson (1992) identified the following design goals for integrated frameworks:

- Support integrated tool sets: The framework should provide support integration at various levels, including that of the user interface, data and the system functionality.
- Support interchangeable tools: The framework should provide

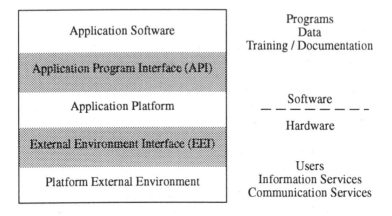

Figure 4.3 The Open Systems Environment (OSE) Model (after Davidson,1992).

efficient interface management.
- Support multiple layers of access: There should not be a single method of accessing the functionality of the software.
- Leverage existing tools: Allowing users to integrate tools already used in a common environment.
- Support other external tools: The framework supports the inclusion of additional, third-party tools.
- Support external computing environment: The tool execution, data exchange and data display should all be designed for a network environment.

Most important to the development of this integration framework is the concept that data integration is transparent to the user. This is by far the most technically difficult task and raises a number of issues that have to be resolved. These include where the data actually resides, what are the most appropriate models for data integration such as the entity-relationship model or the object-oriented model, and who controls the data. This last is more often a political or economic consideration rather than one of purely system design. As Flowerdew (1991) observes, data integration is neither a trivial nor a straightforward process and yet data integration is at the very heart of GIS.

4.7 Corporate solutions

With the move towards system integration across the broad scope of IT in the 1990s, there is a need to develop organisation strategies that use the technology for better decision making. In this context it is possible to identify two trends in the development of GIS. One is well established and based on the generalised GIS product which is identified by its core functionality. The second, and more recent development, is based on the application sector product, which is developed to meet the needs of a particular area such as marketing, property management or environmental assessment.

If the latter trend continues the notion of a single GIS fulfilling all the spatial data handling needs of an organisation is replaced by the need to purchase more than one system within an organisation. These specialist systems can be very precisely tailored to meet an organisation's needs. The resulting organisational models for GIS are shown in Figures 4.4 and 4.5. In one model the generalised GIS is co-located with the basic information resource, such as the base mapping, and acts as a centralised system. In the second model only the data are centrally held, with application GISs located in appropriate departments of the organisation. This model may be further developed by the introduction of localised spatial databases that are subsets of that held centrally.

The importance of this approach is the move towards systems which are designed with integration as the kernel concept. The adoption of the OSE Model for spatial information should lead to the development of a generation of spatial information systems which use new architectures to achieve a far higher degree of system integration.

A corporate GIS solution, as defined by Sheath (1991), has two main functions:

- It allows the disparate information systems to exchange data and users to move between systems.
- It allows the user to perform geographically based enquiries and to have information presented in a graphical form.

It is possible to conceive of the two GIS models in Figures 4.4 and 4.5 existing in an integrated environment where the user either has access to the other systems through the GIS or has access to any system and through that to any other system in a totally transparent way (Figures 4.6 and 4.7). In the former example it is assumed the user has knowledge of the GIS and how it integrates to the other systems. This implies a level of specialist knowledge in the GIS. In the second example, if the user interface is simple and intuitive to use specialist

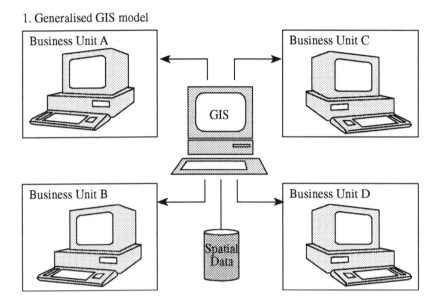

1. Generalised GIS model

Figure 4.4 The generalised GIS model.

skills are not required to undertake basic information query or manipulation. Both these solutions may be regarded as corporate GIS, but the latter, which is based on application GIS integrated with other systems, is more properly a corporate solution.

The key issues that relate to the development of a workable corporate GIS strategy were outlined by Wild (1992, after Margulies and Raia, 1978):

- It is planned.
- It is problem-oriented.
- It reflects a systems approach.
- It is an integral part of the management process.
- It is not a 'quick fix' strategy.
- It focuses on improvement.
- It is action-orientated.
- It is based on sound theory and practice.

Wild concludes that a corporate GIS can provide a powerful

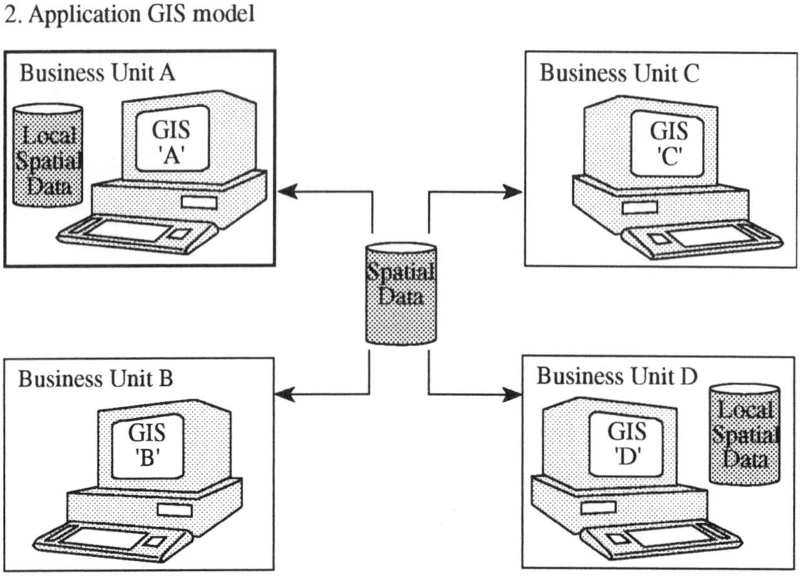

2. Application GIS model

Figure 4.5 The application GIS model.

development tool that can even change cultures and provide new opportunities and new perspectives.

The use of application GIS in an integrated model is likely to lead to GIS being subsumed within broader information systems in the future. This has implications for spatial information and how it is handled. Current levels of user awareness will have to be raised if spatial information handling is to be effective from integrated IS solutions.

4.8 Organisational issues

Such corporate GIS may be used as an 'intervention' to bring about a change in the way an organisation functions (Wild, 1992). Campbell (1991, 1992) identifies three groups of organisational factors that will affect the initial decision to adopt GIS technology and its subsequent implementation and utilisation:

- The organisational context

Corporate GIS solution based on generalised GIS

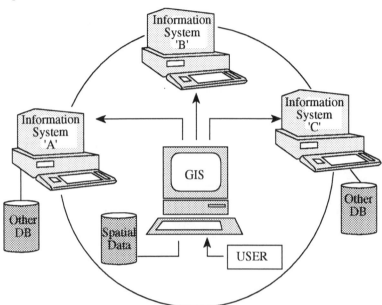

Figure 4.6 Corporate solution for generalised GIS.

- The personalities of the individuals
- The degree of organisational and environmental stability.

The organisation context can be subdivided into internal and external components. The internal components include the organisational structure, administrative arrangements and the identified procedures for decision-making. The complexity of large organisations can make it very difficult to define the precise information handling role of a GIS, particularly in the early stages of evaluating the need and obtaining organisational commitment.

External factors can critically influence the use of any information system. These can be various and are often unpredictable, particularly in their detailed practical implications. They include policy from central and local government, influence of professional bodies, and commercial or economic factors.

Corporate GIS solution based on application GIS

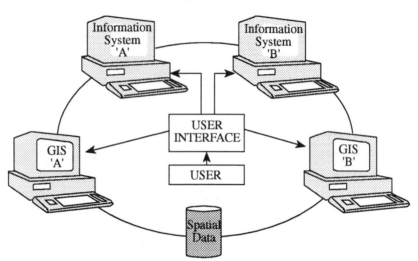

Figure 4.7 Corporate solution for application GIS.

The individuals involved in implementing and running an information system are important in determining its use and effectiveness. Decisions have to be made over who will be responsible for operating and managing the system. How it is perceived by individuals at all levels, either positively or negatively, will determine its acceptance, the emphasis placed on its usage and the level of integration that is achieved.

The issue of stability is of increasing concern. As discussed in Chapter 1, the increasing pace of technological development and the pressure for continuous change leads to organisations constantly reviewing their aims and objectives. As a result, organisational structures and information systems often change and it can be difficult to effect a long term information handling strategy. Where the pace of change is too great this can have a detrimental effect on the effectiveness of an organisation.

Campbell (1992) identifies three conditions for the successful use of a spatial information system:

• The existence of an overall information strategy based on the

needs of users and the resources available within an organisation.
- The personal commitment and participation of individuals at all levels of an organisation.
- A high degree of organisational stability with respect to personnel, administrative structures and environmental considerations.

The importance of these factors should not be underestimated even for stand-alone solutions, but they are of increasing importance for multi-user spatial information systems. The greater the number of users and the wider the impact on the organisation, the greater the influence of these factors.

4.9 Implementation strategies

The implementation of corporate GIS solutions requires medium or long term strategic planning with the acceptance that a return on the high capital investment will only be achieved after a number of years. The implementation process is therefore very important, particularly in an environment where annual budgetary control dominates, such as in local authorities.

Implementation strategies are important, whether developing a multi-million pound corporate solution or a single departmental system. Mounsey and Pearce (1992) identified the following stages in the implementation life cycle:

- Initial introduction of the concept of GIS into the department or organisation.
- A study to establish the potential for GIS in support of the organisation's business.
- System selection to choose the most appropriate system.
- A cycle of implementation and review to evaluate the success of the project and guide future developments.

In any implementation process it is critically important to establish awareness of GIS and its implications for an organisation at all levels from the user to senior management. This is necessary to establish support for a system both in the planning and eventual use. This awareness has to extend beyond the immediate users of spatial information if the system is to be integrated with other information systems.

The implementation process has to address the organisational issues, such as existing information systems, information flows and decision-making hierarchies, and the potential for the GIS to effect change within the overall aims and objectives of the organisation.

Early GIS systems were designed from a very narrow viewpoint, with most attention being paid to the hardware and software. Less effort was committed to issues of data collection, capture and management, which in turn led to failure of systems due to unexpectedly high costs of database creation. Often little thought was given to the organisational issues. More recently the emphasis on organisational efficiency and well-defined business strategies has laid a more solid foundation for integrated GIS. The growth in decision support systems, executive information systems and general purpose management information systems has begun to develop a technology-based information culture among managers, bridging the divide which often still exists between a manager and the technical specialists, such as the engineers or architects.

The essential elements of an implementation process are discussed in detail in various texts, for example Antenucci *et al.* (1991), Aranoff (1989), Clarke (1991) and Huxhold (1991). The problems become more acute when the GIS is being integrated with existing systems. Many implementations try to link a GIS solution into an existing information system where information flow, decision making and organisational structure are optimised for that system and may be unsuitable for the GIS.

The change in emphasis that a GIS may introduce, with greater interest in the use of spatial information, can have profound implications for an organisation. A major part of the implementation cycle for large integrated systems is the identification of threats and benefits which an implementation may bring. This requires considerable understanding of the organisation, in particular its philosophy and culture, in order to effect a smooth and efficient transition.

Failure to adopt a rigorous and well-constructed implementation process will inevitably lead to an unsatisfactory solution. In some cases this may mean the complete failure of the resulting system after only a short time. Complex implementations require support and expertise from within the organisation and advice from specialists with experience in similar circumstances. The effects of success and failure in the implementation process are summarised in Table 4.2.

4.10 Benefits and costs

The implementation of an integrated GIS solution will only occur if the organisation perceives that the benefits outweigh the costs. All implementation processes involve detailed cost/benefit analysis in the

Success		*Failure*
Rigorous	PLANNING	Short lived
Focused	REQUIREMENTS	Diffused
Realistic	APPRAISAL OF EFFORT	Unrealistic
Dedicated, motivated, continuity	STAFFING	High turnover
Adequate financial plan	FUNDING	Inadequate, conjecture
Thoughtful	TIME	Rushed or prolonged
Balanced	EXPECTATIONS	Exaggerated

Table 4.2 The elements of success and failure in a GIS implementation (Antenucci *et al.*, 1991).

planning stage and evaluations of the cost-effectiveness of proposed solutions. The cost/benefit analysis has to be built into a business case that will persuade management to commit resources to GIS above the other demands of the organisation.

For many GIS systems the costs are often easier to define than the benefits. Costs include the following:

- Acquisition of hardware
- Acquisition of software
- Implementation of the system
- Maintenance of the hardware and software
- Data capture or conversion costs
- Staff development costs, including general awareness and specific training costs
- Recruitment of specialist staff

- Consumables and other running overheads
- Site preparation.

Many of these costs are incurred in the implementation process and are relatively easy to define, although some are part of the on-going running costs of the system. The high cost of implementation can be a deterrent and the business case has to justify the spread of the implementation costs over the lifetime of the GIS. This is usually agreed to be between 3 and 5 years.

Data capture costs in particular can be high. Clarke (1991) notes that capture costs can be anywhere between 10 and 1000 times the hardware and software costs. However, with the wider availability of digital data, this may be greatly reduced. The study reported by Capper and Watson (1992) identified data capture costs as being only 38% of the total costs of the implementation (Figure 4.8). The importance of these costs, however, is that the investment in information continues to provide benefits to the organisation long after the original computer system has become obsolete.

The benefits on the other hand can be more difficult to quantify. These may be broken down into three categories (Clarke, 1991):

- Efficiency benefits in reduced duplicated effort, faster data

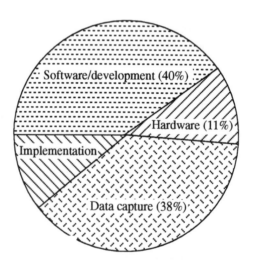

Figure 4.8 Costs of implementing a corporate GIS solution (Capper and Watson, 1992).

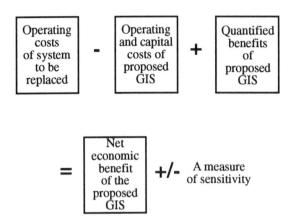

Figure 4.9 Basic economic equation for a cost/benefit study (based on Clarke, 1991).

processing, fewer staff.
- Effectiveness benefits, such as improved quality of decision making.
- Intangible benefits, such improved public image or co-operation between staff.

It is very difficult to put meaningful costs to the effectiveness and intangible benefits, even though these may be very significant. The impact of corporate GIS across an organisation in terms of wider availability of data, improved information flow and a better decision-making structure can be more important than the technical innovation. It is also very difficult to quantify.

Models for preparing a cost/benefit analysis have been proposed based on GIS product values (Dickinson and Calkins, 1988). A precise economic benefit can be calculated (Figure 4.9), which should include a measure of the accuracy or sensitivity of the figures, and then this must be assessed against the intangible benefits and the risks to the organisation.

Capper and Watson (1992) reported on the results of a cost/benefit analysis for a corporate GIS implementation for the London Borough of Barking and Dagenham. Total expenditure was estimated to be in the order of £1 million over a 5 year period. Benefits, including the development of new income sources based on the information available from the GIS, would be in excess of £1.5 million by the end of the same 5 year period, with an annual increase of a further £250,000. The important point was that cumulative costs would be higher than the cumulative benefits until nearly the end of year 3 (Figure 4.10). Thus the business case had to take a medium term view of the overall benefits to the organisation and consider how to fund the initial set-up costs in the first two years. This meant that the implementation would be based on borrowing in the initial phase or be funded from financial reserves.

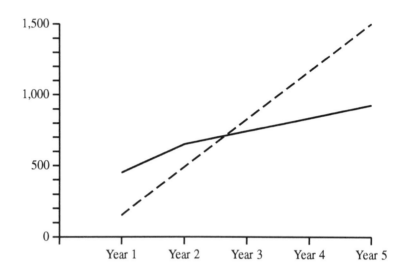

Figure 4.10 Cumulative costs and benefits of a corporate GIS proposal (Capper and Watson, 1992).

References

Antenucci, J. C., Brown, K., Croswell, P. L., Kevany, M. J. and Archer, H. (1991) *Geographic Information systems. A Guide to the technology.* Van Nostrand Reinhold, New York.

Aranoff, S. (1989) *Geographic Information Systems. A management perspective.* WDL Publications, Ottowa

Bracken, I. and Higgs, G. (1990) The role of GIS in data integration for rural environments. *Mapping Awareness,* Vol 4, No 8, pp. 51-6.

Bundock, M. S. and Raper, J. F. (1992) Towards a standard for spatial extensions for SQL. *Proceedings European Geographical Information Systems Conference*, Munich, pp. 287-304.

Campbell, H. (1991) Organisational issues in managing geographic information. In: *Handling Geographic Information*. Eds. Masser, I. and Blakemore, M., Longman, London, pp. 259-82.

Campbell, H. (1992) Organizational and managerial issues in using GIS. In: *Geographic Information. The Yearbook of the Association for Geographic Information*. Eds. Cadoux-Hudson, J. and Heywood, I., Taylor and Francis, London, pp. 337-44.

Capper, B. and Watson, S. (1992) Towards a corporate GIS strategy or Corporate GIS in local government - fact of fiction? *Proceedings Mapping Awareness Conference,* London, pp. 129-39.

Clarke, A. L. (1991) GIS specification, evaluation, and implementation. In: *Geographical Information Systems. Volume 1: Principles.* Eds. Maguire, D. J., Goodchild, M. F. and Rhind, D. W., Longman, London, pp. 477-88.

Davidson, J. V. (1992) Tool integration: Beyond shared files and macros. *Proceedings GIS/LIS,* San Jose, pp. 165-76.

Dickinson, H. J. and Calkins, H. W. (1988) The economic evaluation of implementing a GIS. *International Journal of Geographical Information Systems,* Vol 2, No 4, pp. 307-27.

Flowerdew, R. (1991) Spatial data integration. In: *Geographical Information Systems. Volume 1: Principles.* Eds. Maguire, D. J., Goodchild. M. F. and Rhind, D. W., Longman, London, pp. 375-87.

Huxhold, W. E. (1991) *An Introduction to Urban Geographic Information Systems.* Oxford University Press, Oxford.

Leeson, R. (1992) What is an open mapping system? or a brief swim in alphabet soup. *Proceedings Mapping Awareness Conference,* London, pp. 345-58.

McLaren, R. A. (1990) Establishing a corporate GIS from component datasets - the database issues. *Mapping Awareness,* Vol 4, No 2, pp. 52-60.

McLaren, R. A. and Healey, R. (1992) Corporate harmony - A review of GIS integration tools. *Proceedings Association for Geographic Information Conference,* Birmingham, pp. 1.17.1-5.

Maguire, D. J. (1991) An overview and definition of GIS. In: *Geographical Information Systems. Volume 1: Principles.* Eds. Maguire, D. J., Goodchild, M. F. and Rhind, D. W., Longman, London, pp. 9-20.

Margulies, N. and Raia, A. P. (1978) *Conceptual Foundations of Organizational Development.* McGraw-Hill, New York.

Mounsey, H. and Pearce, N. (1992) GIS selection and implementation - Not simply a question of technology. In: *Geographic Information. The*

Yearbook of the Association for Geographic Information. Eds. Cadoux-Hudson, J. and Heywood, I., Taylor and Francis, London, pp. 368-73.

Sheath, N. (1991) GIS technologies - the issues for corporate systems. *Proceedings Mapping Awareness Conference,* London, pp. 23-32.

Shepard, I. D. H. (1991) Information integration and GIS. In: *Geographical Information Systems. Volume 1: Principles.* Eds. Maguire, D. J., Goodchild, M. F. and Rhind, D. W., Longman, London, pp. 337-60.

Wild, G. (1992) Corporate GIS and organizational development: A local authority perspective. In: *Geographic Information. The Yearbook of the Association for Geographic Information.* Eds. Cadoux-Hudson, J. and Heywood, I., Taylor and Francis, London, pp. 355-61.

5

Low cost spatial information systems

5.1 Low cost strategies

The previous chapter discussed the development of integrated GIS in the context of large organisations and multi-functional systems. There are, however, very many users who are constrained by available resources to using low-end personal computing. This chapter considers the ways in which effective spatial information management may be achieved using low cost strategies.

The common term for the computing platform is the PC or personal computer, but they are also referred to as desktop or micro-computers. The type of configuration ranges from a single, stand-alone machine with its own peripherals and software to large networks of PCs with many users and widely available software and data.

Low cost in the context of this chapter refers to computers whose basic purchase price is of the order of a few thousand US dollars. Total solutions consisting of computer, peripheral equipment and software may be installed for less than US$10,000 and for many practical solutions this can be considerably less. It is currently possible to purchase a computer, laser printer and mapping software with associated database for less than US$3000. Furthermore, while the technology improves prices are being reduced so the cost of such a solution will inevitably become cheaper.

Many organisations have invested limited resources in personal computer facilities over a number of years. There is naturally a reluctance to switch to more powerful computer platforms just for the purposes of GIS, and in many instances it is neither a cost effective approach nor a desirable use of limited resources. Spatial information handling is therefore constrained by existing hardware with investment limited to extra peripherals, software and data.

There is a wide range of software solutions that have been developed for desktop computers, particularly for mapping purposes.

The growth of functionality in such systems as they develop and mature is providing a wide range of solutions that meet many GIS needs.

The greatest limitations in the development of spatial information systems are the costs of human resources, investment in awareness and training, the cost of data and the development of strategies that will deliver cost effective results. These will be discussed in more detail in the following sections.

5.2 Development of the microcomputer

The launch of the first microcomputers in the late 1970s marked a major shift in the availability of digital computer technology. Although early machines were slow with limited memory and poor display capabilities, considerable effort went into improving their performance. This focused on improving memory, increasing the speed of processing, adding hard disk drives for more storage and faster access and the introduction of colour monitors leading to high resolution graphics screens. As a result the power and capabilities of the personal computer have increased considerably and the relative costs have decreased dramatically.

The early microcomputers have not only been replaced by more powerful and more capable machines but have also spawned portable versions and pen-based computers (see section 5.4) which can be used away from a power source. The recent development of a portable Apple Macintosh machine which, on return to the office, may be 'docked' in a housing to which is attached a full screen, keyboard and mouse, is another example. This combines the versatility of the portable solution with the power and capability of the desktop machine.

Such developments point to the future of the personal computer. Increases in PC capabilities are likely to continue, but at the same time there is a move towards integrating personal computers with the telecommunications and audio-visual technologies. This will provide the user with increased flexibility for developing hardware solutions that meet their needs. For example, personal computers may be linked by modem using standard telecommunication networks, fax machines can be connected for the purposes of providing output at a remote location from the computer, and audio compact discs can be played through personal computer CD-ROM players. The integrated television, video recorder, audio and visual compact disc, and computer are available both for the office and the home.

The most sophisticated PC is now as powerful as some contemporary mini and mainframe platforms, making the traditional distinction between groups of machines based on processing power less significant. PCs exist in most workplaces and are common in educational environments and becoming so in the home.

The most important developments for the spatial information community have firstly been the dramatic improvements in the graphics environment (Coll, 1991). Successive developments have seen increases in screen drawing speeds and in colour representation so that 24 bit colour is common on certain platforms. Secondly there has been a corresponding increase in processing speeds to handle the larger volumes of data that enhanced graphics require. By way of example, processing speeds on SUN computers have progressed from 3 MIPS (millions of instructions per second) for the Sun-3/80, to 12.5 MIPS on the SPARCstation 1 and 16 MIPS on the SPARCstation 300.

This surge in the technological developments over the last ten years has in many respects left behind the development of applications. Prior to this application ideas were dependent upon the next leap in computer capabilities before they could be implemented but now technology is running far ahead of the conceptual framework in which such systems operate, both at a software level and at the information management level. The development of new strategies to cope with the rapid advances in personal computers is lagging some way behind.

5.3 Availability of software

The development of desktop computers which act as personal cartographic or spatial information workstations has spawned a wide range of applications software. These products are 'diverse in capability, functionality, price and quality' (Keller and Waters, 1991). The software may be grouped into 7 generic categories:

- Geographical information system
- Thematic cartography
- Computer aided design
- Remote sensing/image management
- Drawing or graphics software
- Digitising packages
- Contouring and surface modelling.

These generic groups reflect the primary use for which the package was designed but in many cases subsequent developments have led to packages being used in more than one area, or changing their emphasis completely. A good example is GIMMS which was originally designed as a thematic mapping package but can now rightly claim to be more of a GIS toolkit.

Appendix B is a summary of the main mapping and GIS related

packages, simplified and updated from that produced by Keller and Waters (1991) and Jones (1992). There are over a hundred different packages which can be used by the spatial information community. Prices for these packages range from a couple of hundred dollars to over US$20,000. Many of the less well known do not achieve a state of maturity and are unable, through a lack of sales, to compete effectively in the market.

The selection of appropriate software presents the uninitiated with a considerable challenge. For most prospective users there may be only a dozen packages that meet their requirements and which they become aware of. Thus the range of GIS and digital mapping software will gradually refine itself as a few widely used solutions become the de facto standards. The pre-eminence of these systems is maintained and enhanced as new users of spatial information automatically turn to these systems. The dominance of the much more mature CAD market by Autocad is an example of this.

The reasons why a system becomes so significant may not be based only on the comprehensive functionality of the software but may be due to other factors, such as the marketing and sales policies, price and customer support. The availability of MAPINFO in a shrink-wrapped box, available as an 'off-the-shelf' product is an example of how the right marketing strategy can increase customer awareness of a particular product.

5.4 Developing technologies

There are two developing technologies which are worthy of mention at this stage. Neither should be regarded as associated only with low cost solutions but they add a significant enhancement to the strategies that may be adopted (see section 5.9).

Pen-based computers are portable computers where the keyboard is replaced by a sensitised screen and pen or stylus. Screens may be laid out like a notepad or form and the pen used much in the same way as a pen and paper. Current versions have capabilities for hand written text recognition which is adequate for most people's hand writing in both printed block letters or script (Hansen, 1991). The systems store data directly into a database which may subsequently be downloaded into another computer.

Current applications of this technology are in the utilities and local government where pen-based computers are replacing paper work orders for service engineers and field workers. In particular, where there is an ageing utility infrastructure and an increasingly competitive commercial environment there is a need to make field crews and service engineers more effective. In the US alone there is estimated to be over 4 million utility workers supporting outside plant and equipment (Hansen, 1991).

Pen-based computers have been used for free form sketches and for digital mapping, in particular in utilities, archaeology and surveying. They are also used in emergency response situations. There is the potential for developing personal GIS workstations, especially as data storage and display technologies improve. The result should be the ability to hold subsets of GIS databases on pen-based computers for field update with both real-time communication links and batch downloading back at the office as shown in Figure 5.1.

An important consideration will be the development of appropriate software and user interfaces that are 'pencentric' or based on a pen. The ability to use both the versatility of a windows approach with menus and icons together with the ability to select when to write responses or notes makes the pen-based computer potentially much more flexible. Issues relating to user interfaces are discussed in more detail in Chapter 9.

The software is also important and can not simply be microcomputer

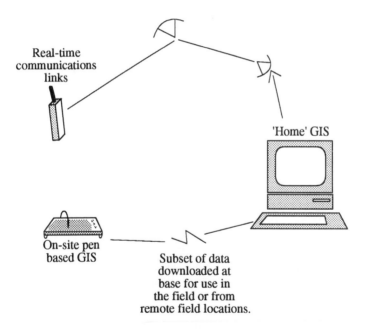

Figure 5.1 The use of pen-based computers for GIS.

software 'shrunk' to fit within tighter technical constraints of the portable machine. Wilson (1991) notes that the software will be built around the completely new mode of human-computer interaction, one that is particularly well suited to GIS.

The result should be the ability to hold subsets of GIS databases on pen-based computers for field update with both real-time communication links and batch downloading back at the office as shown in Figure 5.1.

The other technology worthy of a mention here is Global Positioning Satellites (GPS). GPS really serves two markets. Firstly it is used to create geodetic frameworks which are used in topographic mapping. This is not a low cost application. Cain (1991) notes how the use of GPS for generating continuous geodetic control frameworks to underpin GIS projects has brought about co-operation between state, county and city government in the US.

The second is in the provision of GIS attribute information, which can be achieved using low cost, low resolution GPS technology (Cross, 1991). GPS can be used to determine position in the field for those collecting attribute data for GIS. Small portable receivers can give positional accuracy to within 100 metres and the use of two receivers where one has a known position, called relative positioning, can improve the accuracy to around two metres (Buchanan, 1992). Linked to pen-based or other portable computers, GPS can provide an appropriate strategy for many data collection activities.

There is also growing interest in the use of GPS for tracking moving objects, such as vehicles. Goad (1991) details how GPS can be used to track a moving vehicle for the purposes of generating an accurate road map. Various applications in emergency planning and vehicle navigation have also been developed.

5.5 Stand alone solutions

The typical microcomputer workstation for spatial information consists of a PC, input device and output devices (Figure 5.2). All PC solutions can be considered to be desktop machines with small and generally decreasing footprints, or desk area that they cover.

The input device is usually a digitising table which can range in size from an A4 graphics tablet to A3 table digitisers and up to A0 for floor standing digitisers. Even the table digitisers require more space than most people expect and need to have accompanying layout space for maps and papers. The A0 digitisers need a lot of space to accommodate the table, which can usually rotate into a horizontal position. They have to be located adjacent to the PC in order to be able to see on screen digitising and commands.

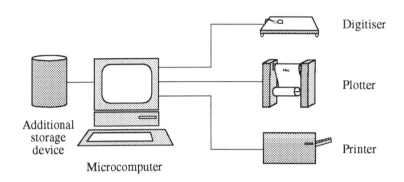

Figure 5.2 Configuration of a typical microcomputer workstation.

Typically such solutions require two output devices, one for text and the other for images. These can range from dot matrix to laser printers and from pen plotters to colour electrostatic plotters. The quality of the output is a major factor in establishing the system and often warrants relatively high levels of investment to establish a resource that will meet the needs of the users for some time to come.

Many plotters will cost as much if not more than the microcomputer. In particular the size of map output has to be addressed. It is possible to purchase A3 pen plotters that take up little desktop space but A1 or larger plotters require considerable room. They do not, however, have to be co-located with the computer.

The space needed for a stand alone system can vary from a single desk to an area in excess of 6 square metres. The amount of space required to lay out maps and diagrams is often forgotten when planning the ergonomics of a GIS set-up.

The choices for configuring the solution are almost limitless. Various levels of computer processing power, coupled with peripheral devices, can be tailored to meet the individual needs. A wide range of additional storage devices can also be added, such as extra hard disk storage, CD-ROM or optical drive.

The costs of setting up a stand alone system at the bottom of the range can be as little as US$3000. This will greatly depend on the software required and its computing requirements. Many of the more sophisticated packages not only cost more but require enhanced memory capabilities and work better on the faster machines. This can add to the cost of the computer.

While very low cost solutions may be adequate for a user's needs, particularly for one-off tasks which are clearly defined, they may have a limited life as a general purpose tool. Some extra initial investment may pay dividends in the longer term, particularly in the provision of more suitable software and better output devices.

5.6 Network solutions

One of the advantages of starting with a small stand alone system is the ability to upgrade the system by linking it into an existing network or by adding microcomputer workstations that can be networked together.

A review of the basic types of network, star, tree, ring and linear or bus network was given in Chapter 4 (after Antenucci *et al.*, 1991). Microcomputers may be used to establish complex and very effective network solutions, often being integrated with more powerful workstations and minis. For low cost spatial information systems a ring network is most often used, where each device has some form of intelligence and can receive and send messages.

One reason for upgrading two or more stand alone systems to a network is to share data, since duplicating large spatial databases incurs costs. Thus a common model is to establish a single database for commonly used files such as the topographic maps. All other aspects of the systems are stand alone, with the necessity to purchase multiple copies of GIS software (Figure 5.3). Users who wish to use more than one software package, such as a GIS, word processor or spreadsheet, must have all the necessary applications resident on their machine. Each machine must have the capability of running the resident software and of storing temporary data files.

This solution may be considered suitable where each user of a computer has a closely defined task requiring constant use of one of more applications packages. The software is dedicated to that user and they create their own data which are not shared until they are deposited with the other accessible data. The size of shared data area need only be as large as the expected database and only limited security and management controls are necessary.

Where the users have a less rigorously defined task and require access to various systems at different times or need to access data from different users for further processing, then an alternative network strategy is required. Figure 5.4 shows a network where all data is stored in a central database and all software is resident on the network server. The networked versions of the software are accessible by any of the workstations, usually up to the number of licences purchased. For

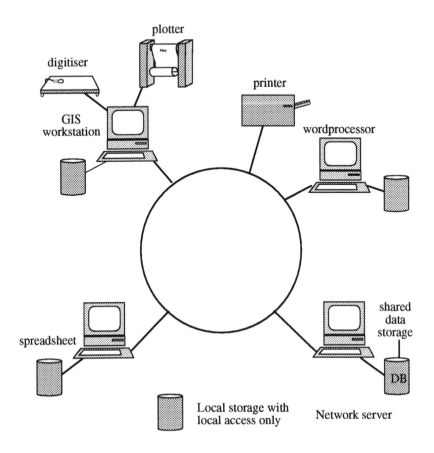

Figure 5.3 Network model for sharing data.

example in a teaching environment it will be necessary to have enough licences to allow each student access to the software at the same time.

While this model has the advantage of preventing software redundancy where copies of packages lie idle for large proportions of the day and imposes stricter control on the data, there are also limitations. Network software is cheaper than multiple stand alone copies but upgrading from a stand alone copy to a network version usually incurs extra cost. If only a few licences are purchased there may be times when users are denied access to software. Also many of the GIS and digital mapping packages are not available in network versions,

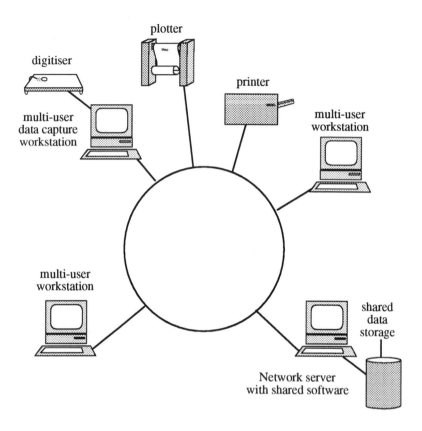

Figure 5.4 Network model for shared software and data.

particularly the lower cost systems.

Much more rigorous management and security controls have to be put in place to ensure data are not lost or inadvertently changed and that data being used by one individual are only accessible by another under strict conditions. Most importantly for many users the responsibility for maintaining copies of data falls to an appointed individual, such as the network manager. The shared data storage is likely to be larger, with considerable space required for temporary files, a common feature of processing in many GIS systems.

It is possible to combine these two models so that users have the choice of software either run from the local machine or over the network and can select to store data either locally or on the server.

In both these network models output peripherals can be placed on the network for access by any user. This clearly reduces the need for multiple peripherals and saves costs. In the first model, data capture must be linked to the GIS workstation, while in the second model, general purpose data capture workstations can be established.

5.7 Data availability

No matter what the computer platform on which the spatial information system is operating, the availability of data remains the most significant problem. The key issues relating to data discussed in Chapter 2, namely capture, ownership, maintenance, security and dissemination, are all equally as important to users of PC solutions.

For those running low cost spatial information solutions data can become a significantly higher proportion of the set-up and/or the operating budget. This is because the purchase of third party data or the capture of data in-house is not significantly cheaper just because the platform and software cost less.

In the UK, for example, there is one standard charge for a large scale map unit, and in fact additional costs are incurred if the user wishes the data in a DXF format, something which might be expected of many PC users. The only difference in costs relates to the storage media, which is higher for tapes than floppy disks.

The capture of data in-house for those with low cost solutions may also incur greater costs where small A3 or similar digitisers are being used. Extra time is involved either in moving maps about a digitiser with a small active area or in creating many small map blocks which subsequently have to be joined together with all the editing this might entail. Efficient digitising of most standard maps requires the use of an AO size digitising table, but the cost of these may be twice or three times that of the computer platform.

Maintenance of data and security are simpler to manage in stand alone solutions where an individual is responsible for the system and its usage. They remain important but cost little. It is more difficult to manage when there are multiple users of a single platform, each using the same data but for different purposes. Maintenance of current versions of the base mapping, back-up files and current derived map products need to be handled properly. The problems of a data processing failure due to insufficient disk space are a common experience and can lead to indiscriminate removal of files. In more complex PC networks, the data management issues are similar to those in larger multi-user systems.

The dissemination of information from a stand alone solution is by the physical transfer of files using floppy disks or other storage media or in hardcopy output. For many small organisations, this approach is perfectly adequate. Costs are limited to the type and quality of output device required.

The efficiency of this approach is determined by the volume of output required and the number of users wishing to access the output device. It can become problematic when multiple copies of a map are required if the available device is slow. Problems may also arise when output can not be queued on the network leading to a need to constantly resubmit a printing job.

The network setup must also consider efficiency of computer usage and operator practices. If the printer setup is such that the computer from which the output was sent is unavailable for other tasks during the printing process this is a very inefficient use of both computer and human resources.

5.8 Human resourcing issues

The problem of staffing spatial information systems in small organisations is based around the need for individuals to have skills in more than one area of expertise. Stand alone solutions or small networks will usually be located in organisations or departments where there are no computer specialists. The implementation of a GIS or mapping system will usually fall to one or two individuals who initiate the procurement process based on their own enthusiasm. This may be based on perceived enhancements to their existing tasks, the prospect of new avenues of work or simply the excitement of using graphics-based technology.

The enthusiasm of the individual is extremely important as it often drives the procurement process, persuading senior staff to invest in the resources and encouraging colleagues to take an active role. Increasing staff awareness often falls to the enthusiast, who in turn is often limited by the amount of formal training available both in the general concepts and in the selected system. Many desktop solutions are learnt by people spending hours of their own time working through system tutorials and manuals.

The enthusiasts in an organisation are usually professionals in other fields, such as town planners, engineers and ecologists. They are able to combine effectively their understanding of their subject and the use they make of spatial data with the potential of GIS. However, a lack of knowledge about computers and often only limited practical experience can severely hinder both the procurement process and the implementation. It is easy to envisage the results of sophisticated processes, such as complex overlay operations, without appreciating the intervening stages necessary to reach the end objective. As a consequence it is easy to purchase a system that does not meet the needs of the organisation and for the proposed tasks to take far longer than originally planned.

There are problems for the organisation in investing all the GIS skills in one or two individuals, who at the same time function in other areas. This can lead to inefficient use both of the system and of the individual's time, particularly in the early stages.

It is tempting to anticipate such problems from the organisation's point of view and to retain elements of the old manual system alongside a GIS or digital mapping system. This may be necessary in the short term to complete existing tasks, but may work against the successful implementation in the medium and long term. This may be reflected in less staff commitment to change and will not encourage a dynamic and diverse approach to problem solving based around the new technology.

The upgrade of a stand alone system to some form of network requires a higher level of computer knowledge, often not available amongst existing staff. This implies an investment in greater staff resources and often represents a major hurdle to be overcome for many small organisations. Such human resourcing issues can effectively constrain the development of integrated solutions.

5.9 Implementation issues

In Chapter 4 a range of implementation strategies was identified for establishing integrated spatial information systems. These may be equally applied to low cost solutions, but it is important to highlight those issues which are of particular concern.

Cost is always a very important factor in the establishment of an information system, but for low cost strategies a cost ceiling has been imposed which is usually both rigorous and demanding of those with the task of establishing an effective system.

Implementation time-frames for low cost solutions are either very short since little money is available to spend on consultation and preparation, or are extended because an individual has to find the time to progress the implementation while also fulfilling other job requirements. The effects can be that an implementation is too hasty and ill conceived or conversely there is a loss of interest in the proposals as little is seen to materialise.

Data as we have discussed are a fundamental problem in which costs of collection and/ or purchase have to be met. The availability of resources for establishing a working database have to be carefully considered and an appropriate plan established that links the database creation into the needs of the organisation in both the short and medium terms. Where costs are considered to be paramount, appropriate accuracy and quality measures need to be defined that meet the task requirements but do not overburden the data capture process.

Low cost GIS strategies often use locally available hardware and therefore the range of suitable software may be limited. The selected software solution may have as much to do with price and availability as with functionality. It is important to identify possible upgrade paths to increase the capabilities of the system.

The integrated nature of spatial data usage should also be addressed for low cost solutions. Organisational benefits may be derived from the use of a suitable solution with longer term cost savings and efficiency improvements. These are often the most difficult to justify for low cost systems.

5.10 Low cost GIS users

The range of users who are considering the use of spatial information is increasing. The availability of low cost alternatives has prompted the consideration of GIS by a wide range of smaller organisations, companies and local and central government departments. This is true in North America, Europe and the rest of the world.

The effect of this growth in interest is the development of an enormous potential market for spatial information management, particularly in the area of land information systems, environmental monitoring and management, urban growth and renewal and health care. The constraints on the growth of low cost GIS are principally data availability and awareness and education.

An example of a low cost GIS implementation is that for Bassendean Town Council, Western Australia (Devenish, 1992). This was established to generate information necessary for town planning decision making. The GIS was used to compile information on planning criteria such as current landuse, existing housing densities, condition of housing stock and reticulated sewerage provision. The use of PC-based solutions was considered most suitable for the Bassendean Town Council as there are only limited users who directly access the information. The total costs of establishing the GIS for a initial project was A$31,426, including all hardware, software, data, training, data collection for the project and consumables. This compared very favourably to the estimate of A$15,000 for the production of a single landuse map without the assistance of a GIS (Devenish, 1992).

Low cost solutions are of particular interest to many organisations that do now have access to more sophisticated technology or funds for its purchase. The problems of GIS usage in the Commonwealth of Independent States (CIS), formerly the USSR, is graphically demonstrated by Koshkarev (1992) in which he notes that lack of education and awareness in GIS, the shortage of experts and the problems of data availability make it difficult to progress even PC-based systems that have been made available under forms of technical aid from the US. There are, however, some GIS systems that have been developed in the CIS countries and these gain from the fact that they were developed to run on low order computers. The importance of PC-based systems for the growth of GIS has also been noted in a countries such as Hungary (Kakonyi, 1992) and India (Krishnayya, 1992).

References

Antenucci, J. C., Brown, K., Croswell, P. L., Kevany, M. J. and Archer, H. (1991) *Geographic Information Systems. A Guide to the Technology.* Reinhold, New York.

Buchanan, H. J. (1992) One man ground survey. In: *Geographic Information. The Yearbook of the Association for Geographic Information.* Eds. Cadoux-Hudson, J. and Heywood, D. I., Taylor and Francis, London, pp. 262-7.

Cain, J. D. (1991) GPS to control counties and municipalities. *Proceedings GIS/LIS,* Atlanta, pp. 247-56.

Coll, D. C. (1991) Developments in equipment and techniques: Microcomputer graphics environment. In: *Geographic Information Systems. The Microcomputer and Modern Cartography.* Ed. Taylor, D. R. F., Pergamon Press, Oxford, pp. 39-59.

Cross, P. A. (1991) GPS for GISs. *Mapping Awareness,* Vol 5, No 10, pp. 30-4.

Devenish, S. (1992) Low cost GIS implementation on a PC. *Proceeding Mapping Awareness Conference,* London, pp. 117-28.

Goad, C. C. (1991) Positioning with the Global Positioning Systems Satellites. *Proceedings GIS/LIS,* Atlanta, pp. 240-6.

Hansen, C. A. (1991) Using pen-based computers for GIS. *Proceedings GIS/LIS,* Atlanta, pp. 964-72.

Jones, K. (1992) A comparative survey of PC-GIS for urban and regional planning. *Proceedings Mapping Awareness Conference,* London, pp. 7-16.

Kakonyi, G. (1992) International GIS: Hungary. In: *1993 International GIS Sourcebook.* Ed. Eynon, D., GIS World, Fort Collins, pp. 282-3.

Keller, C. P. and Waters, N. M. (1991) Mapping software for microcomputers. In: *Geographic Information Systems. The Microcomputer and Modern Cartography,* Ed. Taylor, D. R. F., Pergamon Press, Oxford, pp. 97-128.

Koshkarev, A. (1992) Geographical Information Systems in the CIS: A critical review on the critical state. In: *Geographic Information. The Yearbook of the Association for Geographic Information.* Eds. Cadoux-Hudson, J. and Heywood, D. I., Taylor and Francis, London, pp. 35-40.

Krishnayya, J. G. (1992) International GIS: India. In: *1993 International GIS Sourcebook,* Ed. Eynon, D., GIS World, Fort Collins, pp. 284-5.

Wilson, P. M. (1991) Taking it to the streets: GIS on handheld computers. *Proceedings GIS/LIS,* Atlanta, pp. 587-92.

6

Copyright and legal issues

6.1 Who owns the data?

A major limitation on the development of spatial information systems has been the availability of data. Without suitable data, particularly digital maps, the effective use of GIS is not possible.

Many countries have begun the process of converting their paper map records to digital form, with some nearing completion. The UK's Ordnance Survey will complete the digitising of the large scale 1:1250 and 1:2500 series (some 200,000 map sheets), which cover most of the country, by 1995. Some countries are equally advanced in the creation of a national spatial database while others have either not begun the process or do not have current mapping of the appropriate scales necessary to meet the needs of a burgeoning GIS industry.

Where maps are not available from the national survey agencies then international organisations or commercial firms become involved in establishing a current map base. Even where large scale map cover exists there may be good reasons for the development of alternative map databases. The Automobile Association in the UK, together with European partners, have established their own digital road network for the continent (Robbins, 1992).

The problem for most GIS users is knowing what the availability of data is, who owns the data and what rights they have to use those data. There is also the issue of integrating datasets for the purposes of analysis (a subject we return to in Chapter 8) and what rights exist for passing such derived data on to third parties.

The problems of data ownership and copyright also lead into the vexed questions of ownership of a 'value added' dataset and to legal issues of responsibility for data accuracy, currency and use. There are also legal issues relating to liability of GIS systems and who is responsible for quality of the results produced. There are cases where courts have decided that liability for errors has been, alternatively, with the data supplier, the system developer and the information user.

6.2 Law related to information

Information in general, and land-related information in particular, are sought after for a variety of reasons. Some are collected to fulfil legal mandates, such as the requirement imposed on local authorities by the UK government to collect information on street works as a result of the New Roads and Street Works Act 1991 (Wallwork and Goodwin, 1992). Other requirements come from a demand for public information or from organisations involved in commercial enterprises. The result is that information has a set of attributes which distinguish it from other commodities and make it difficult to develop rules which govern its collection and use.

Epstein (1991) noted that 'data and information are distinguishable from all of the other commodities' which are subject to the legal system, and identified the following, which make information peculiar from a legal perspective:

- It is hard to measure the quantity of information.
- It is difficult to define and implement property rights (i.e. appropriability) in information. Issues of ownership, access, sharing, privacy and political control are important. Suppliers will not provide information if there is over-use and no economic return.
- Information is not depleted by use, unlike a car or pair of shoes.
- Inspection of a product is characteristic of a commodity market but inspection of information is often tantamount to use.
- Information can be transmitted and disseminated at low cost, so value generated from the delivery process are negated.
- Information has the character of the public good. It is often generated at public expense to satisfy a legal or administrative mandate. How it is used after being generated is not always clear.

The legal framework within which information is controlled has to take into account the increasing desire to capture and disseminate information both as part of the function of government and for commercial gain. The increased capabilities of computer-based information handling systems results in the transfer of data and information more easily, faster and in ever increasing volumes. Technological change is also increasing the ability to obtain data and information, either with the owner's approval or without. It is difficult to establish legal systems that make data more widely available without penalising the data providers.

Inevitably, legal statutes take time to prepare and put into place. The result in many countries is that the legal framework in which information transactions are conducted was not designed to meet the current needs. Existing legislation may or may not cope with aspects of the information revolution in an appropriate manner and it may be that the inadequacies of the system take time to be recognised, particularly where legal precedent needs to be established. The result is a legal framework which is somewhat out of step with a rapidly developing technology and increasingly information-based society.

6.3 Access to information

Inevitably the law relating to data ownership and the use of data within information systems is complex and treated in various ways in different countries. In some there is the concept that information is essential to the establishment and maintenance of a 'free' society where ideas and views can be developed and expressed. In the United States this is enshrined in the First Amendment to the Constitution.

This is by no means true throughout the world. In many countries much information, particularly that collected by central government, is carefully controlled and limited in its distribution. In many countries the topographic maps produced by government agencies are not available for public use.

In some countries access to information collected by central government is enshrined in law. The US Freedom of Information Act (1966) requires that 'each agency, on request for identifiable records made in accordance with published rules ... shall make the records promptly available to any person' (Epstein, 1991). There are also similar laws in most of the states in the US.

There are restrictions placed on the access to data by such legislation which means that not all information is freely available. However, usually the burden lies with the government agency to prove the data are excepted from the legislation and not with the individual to prove the data are covered by the legislation. The sorts of restrictions include data specifically requested to be kept secret for reasons of national security or foreign policy, data related to personnel that could be regarded as an invasion of privacy, trade secrets and commercial or financial information which is privileged or confidential. One specific restriction detailed in the US Freedom of Information Act is geological and geophysical information and data, including maps, concerning wells (Epstein, 1991).

The European Commission has proposed a freedom of access to information to ensure dissemination of information throughout the community (CEC, 1990). The proposal guarantees individuals the right to information about the environment which is in the possession of public authorities, supplied at a reasonable cost. This includes all existing data collected or prepared in written document form, in databases or visual records. It also includes certain information supplied by others to government. It is not clear what is meant by environmental data in this case but it would certainly seem to include topographical information (Rhind, 1992a).

There are also further proposals for European Community legislation that, in the event of a dispute between a data producer and user, will place the burden of proof on the data supplier. The data supplier will have to prove the user is at fault (Dale, 1991).

The importance of statutes like the Freedom of Information Act is to enshrine in law the principle of free flow of information on the premise that information is a commodity in the free flow of commerce. However, the complexities of the legal framework surrounding information increase as information becomes more distributed. As Antenucci *et al.* (1991) observe, the more that institutions share information to improve the benefits-to-costs ratio, the more complex the legal and institutional issues become.

Conflicts on access to information often arise from the desire to use government data by commercial organisations who view such data as a public resource. Government structures for the dissemination of information and for controlling its subsequent use are often unsuited to information management in a technological environment.

6.4 Data ownership and copyright

Laws relating to the ownership of spatial data are mainly based on copyright but can include legislation on confidentiality of information and commercial contract laws.

Copyright is distinct from laws such as the Freedom of Information Act and is designed to protect the commercial value of a work, including databases. Copyright covers the products or information generated by an individual or organisation. The copyright laws protect the producer of a product or information from having it copied without their permission, thereby preventing other individuals or organisations from making a profit out of someone else's work.

It is not possible to copyright an idea, only the product which results from that idea. For example an idea for an innovative way of representing information is not copyrightable, but any software developed to implement that idea is covered by copyright. It is possible that government databases may be copyrighted but still subject to disclosure under national freedom of information legislation.

The relationship between owners and users of geographic information gives rise to a number of complex legal issues. These include the intellectual property rights which may exist in geographic data, what protection is enjoyed by the owners of those rights and how those rights may effect potential users (AGI, 1992a). Intellectual property rights cover the rights of an individual or organisation to claim copyright of a piece of information because of the knowledge and expertise invested in collecting or creating that piece of information. Intellectual property rights also includes patents, trademarks, proprietary technology and trade secrets (Smith and Parr, 1989).

In the UK, the Association for Geographic Information established a working party to look at the complex issues as they relate to spatial data in order to help clarify areas of concern to both suppliers and users of geographic information (AGI, 1992a). A number of UK laws relate to the protection and use of spatial databases:

- Copyright, Designs and Patents Act 1988
- Data Protection Act 1984
- Computer Misuse Act 1990
- Law of confidentiality, which is largely based on legal judgements
- Contract Law

The most important of these is the Copyright, Designs and Patents Act 1988, which specifically protects certain types of work, including literary works, artistic works (which encompasses maps) and cable programmes or any item in a service by which information is sent by means of a telecommunications system, which includes a computer. To be included a work must be recorded in some form, such as on paper, tape or computer disk.

A database does not attract copyright in its own right but individual elements such as each separate piece of data and the computer program for loading, retrieving or manipulating the data do. If the totality of the database meets the requirement that it is 'original' then it will also be the subject of copyright. However, the degree of originality required to qualify for copyright varies from country to country and has led to conflict. In the UK a map whose detail is derived from a satellite image but which uses the Ordnance Survey map for interpretation of features is liable to have infringed the copyright of the Ordnance Survey map.

The copyright owner is usually either the author or creator or, where they are employed, the company. The duration of copyright varies but is typically 50 years either from the death of the author (who is not necessarily the copyright holder) or from the time a computer-generated work was made.

Permission must be sought of the copyright holder if any form of copying of the work is to be undertaken, including copying from floppy disks to hard disks or memory, issuing of copies to the public, presenting the work in public and adapting a work, such as translating a computer program.

Copyright is also protected under international law. In particular the Berne Convention and the Universal Copyright Convention mean that works created in all industrialised countries will attract copyright in the UK.

6.5 Added value information

The complexity of the law relating to issues of ownership and copyright have prompted much debate in the spatial information community on the question of 'added value data'. For example, if a local authority purchases the right to use two datasets from different sources and combines these using their own information to create a third dataset, who is the copyright holder of the third dataset (Figure 6.1)?

The problem does not exist for those who wish to create, manipulate and display their own information. Nor is it a problem for those who wish to purchase data simply for the purposes of manipulating and displaying that data set. But problems arise when organisations wish to create, manipulate and display data that are derived from multiple sources. This is likely to be a growing problem, since the power of GIS is the analysis of multiple data sets. The introduction of modelling and the question of how these varying datasets are combined affects the

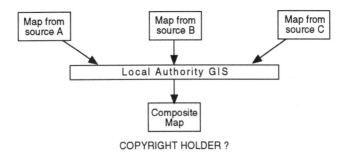

Figure 6.1 Problems of identifying the copyright holder for 'value added data'.

outcome. Thus intellectual property rights are important as well as the rights of the original data owners.

This type of problem should be resolved under the terms of the contract by which a person or organisation buys the right to use the data. However, in cases of dispute, each will be decided on its merits, but for most information managers who need to address issues of copyright it will be necessary to seek legal advice.

6.6 Liability and GIS

Another aspect which has a significant impact on the development of spatial informations systems is the area of liability. A major concern is the uncertainty that results from reliance upon information which is later shown to be inappropriate for the use for which it was intended (Epstein and Roitman, 1990). Liability can be divided into two broad areas (Epstein, 1991):

- *Contractual liability.* An individual's or organisation's liability is determined by an agreement between parties which should clearly set out duties and responsibilities. If this is properly done, failure to meet those obligations by any party to an agreement can lead to legal action by other parties to the agreement who wish to seek some form of redress.

  In some cases a written contract is not required since certain rights are already enshrined in the law. For example, a buyer of a product has certain rights and is entitled to expect the product to work as specified. In the case of complex computer installations or the creation of databases, detailed contracts need to specify what the buyer expects and what are the limits of the seller's liability.

- *Negligence.* This arises when an individual or organisation fails to exercise 'reasonable care' in undertaking a task which leads to some form of damage to others. The definition of 'reasonable care' relates to an expected level of ability applied to a particular person or task, for example an electrician would be considered negligent if he incorrectly wired a plug, leading to damage of an electrical appliance or fire, or to personal injury.

Contractual liability is relatively straightforward to define. Legal interpretations are based on the agreed documents and often concern aspects of interpretation. Clarity in drafting a contract, with time taken to ensure all parties are fully aware of their obligations, can limit disputes later. Problems can arise in areas where contracts are not normally used, particularly in areas of public sector information management.

Negligence is a more complex area, since legal judgements are based on case law, a view of what constitutes reasonable care for a particular task and the circumstances pertaining to specific cases. The recent emergence of spatial information systems means that little case law exists to help resolve disputes.

When considering what constitutes reasonable care it is clearly unreasonable to expect error-free work in the majority of circumstances and acceptable levels of operation need to be defined. These may well form part of a contractual undertaking. In the case of a GIS these may include data access times, display parameters, security controls, backup facilities in the case of system failure and acceptable system down times.

The area of liability extends to hardware and software suppliers as well as data suppliers. Each will be responsible for their individual products but 'may also incur part of the overall responsibility for the satisfactory workings of the whole hardware-software-data-human interface system' (Rhind, 1992a). Courts may find themselves in the position of apportioning the blame between the various contributing suppliers.

6.7 Accuracy of information

Most of the existing case law that relates to GIS is centred on the accuracy of maps or map-based information. In the US a number of cases have come to court in which the producer of information has been found 'liable for fatal injuries caused, in part, by incorrect or non-standard data' (Aranoff, 1989).

Aranoff (1989) identifies four broad areas in which users of GIS have a responsibility to ensure minimum standards of quality for their information:

- *Accuracy of content.* Those involved in compiling maps or spatial databases have to consider if it contains all the necessary information appropriate to its use and to ensure that the data are correctly located. In 1978 the United States Federal Government was found to be negligent for incorrectly showing a broadcasting tower on an aeronautical chart, which led in part to a fatal aeroplane crash.
- *Accuracy of context.* Interpretations of information from maps are based on the relative positions and the geographical context of the features shown. This has to be kept in mind when compiling maps which will be used by individuals not involved in their compilation. The map compiler could be regarded as

negligent if the user could reasonably draw an incorrect conclusion. In 1981 a US company who produce aeronautical charts was found to have contributed to a fatal aircraft crash by using a non-standard perspective for an airport approach chart which confused the pilot.

* *Data format.* The format of information, particularly information required by statute to be compiled by local or central government, is very important to data access. Changing formats can affect an individual's or organisation's access rights to those data. In the UK, local governments are legally required to maintain a map database of all public footpaths and bridleways. While it might be reasonable to recompile the data in a GIS this would require a change in the existing law to permit a digital data format.

* *Combining of data sets.* The issue of negligence is also important when combining different datasets. Problems arise in determining the accuracy of a dataset derived from a combination of data from various sources. Careful quality control procedures must be set up to ensure consistent accuracy levels are achieved, even though this can be extremely complex when combining multiple datasets.

Failure to give appropriate consideration to data quality issues can be regarded as negligence. The problem for the spatial information manager lies in defining the 'acceptable minimum standard'. The development of data content and data accuracy standards and the adherence to these standards is a way of overcoming some of the problems, but individuals will still be expected to use their professional judgement in exercising reasonable care in the creation of a product.

6.8 Inappropriate map usage

While a map product may have no significant errors or omissions and the design of the map has addressed all the issues related to context, there remains the problem of using maps or other spatial data products for unintended purposes or in unintended ways. A hill walker who gets lost can not reasonably be expected to blame the map if he is using a 1:250,000 road map.

There is an expectation that users of maps or other spatial information will use their professional judgement in evaluating information and will not rely blindly on what the computer or a map shows them. A map symbol of a church incorrectly placed in the sea should be easily identified by the map user as a gross error.

A case in the United States relating to land ownership and the position of an ordinary high water mark for a lake concluded that the state was liable for harm imposed on a landowner resulting from the use by the state of a map for a purpose for which it was not intended (Epstein, 1987). These types of problems are likely to increase with the wider availability of mapping through GIS (Rhind, 1992a).

Increased awareness of how maps are compiled and what they show will be necessary to limit this type of problem. In particular accuracy, currency and appropriateness of specific maps for specific applications need to be understood. There is an onus on the spatial information community to ensure greater understanding of GIS products to ensure their effective use.

6.9 Data availability and government policy

Many of the problems currently existing within the GIS community stem not from who owns data but from how data can be used and what charges are made for them. This latter is particularly important since the cost of data can account for such a high proportion of the running costs of a GIS.

In the US only modest costs covering copying and media charges are levied on government information under the Freedom of Information Act. Maps produced by the US Geological survey (USGS) are freely available for use within the USGS and for certain federal activities, and for others the charges only cover the costs of reproduction and distribution (Rhind, 1992a). The same is also true for digital data, a policy strongly supported by the private sector who are able to sell added-value products at commercial rates. Key to this policy is that there are no copyright restrictions on the use of government data, which may be copied to as many users as required without permission or payment (Rhind, 1992a).

In the UK, on the other hand, it has been government policy for some time to try and recover part or all of the costs involved in compiling and maintaining the country's map base. The Ordnance Survey achieved a 68% cost recovery in 1991/92 (Rhind, 1992b). Thus users of UK maps produced by the Ordnance Survey are faced with much higher charges than for comparable data in the US.

It has been argued that central government policy in the provision of spatial data has a lot to do with the success of the GIS industry. Estimates from the US suggest that the wider availability of map data has spawned a relatively larger GIS industry than in the UK, leading to longer term benefits for the economy and increased revenues for the central exchequer. The UK Government policy can therefore be considered as restrictive, leading to short term returns on the costs of producing maps but without the longer term growth in GIS needed to sustain the industry (AGI, 1992b).

The availability of a national map base compiled over many years and maintained by the state is viewed by many as a public asset. In this regard, responsibility for maintaining the data is that of government but access to the data should be available to all potential users. However, the issue of whether there should be charges for geographic data produced by central government relate to government policy.

The other issue is what level those charges should be. Rhind (1992b) identified a number of factors that determine the prices charged by the Ordnance Survey:

- The nature and magnitude of the task required to bring a product to market;
- OS costs and efficiency in carrying out the necessary work;
- The anticipated market size;
- Marketing policy decisions (within the strict constraints that do not permit cross-subsidisation between map series);
- The need for equity of treatment of customers, consistent with the OS position as a government department;
- External OS factors such as UK government policy and European Community directives which are incorporated into UK law. The former includes the need to generate sufficient funds for maintenance of the asset and avoidance of monopoly trading practices;
- The need to ensure prices are as low as possible overall to maximise use of the asset;
- The desirability of future pricing policy being reasonably predictable;
- The impact of loss of revenue due to unauthorised uses.

The problem of high levels of cost recovery imposed on the users is that it encourages infringement of copyright laws. The impact on revenue by unauthorised use of OS data is considered to be a growing problem. Government policy on charging is therefore very closely linked to the issue of data ownership and data usage.

References

AGI (1992a) *Report by the Copyright Committee, The Association for Geographic Information to the AGI Conference 1992.* AGI, London.
AGI (1992b) *Final report and recommendations to the AGI Council by t*

he Association for Geographic Information Ordnance Survey Charging Round-Table. AGI, London

Antenucci, J. C., Brown, K., Croswell, P. L., Kevany, M. J and Archer, H. (1991) *Geographic Information Systems. A Guide to the Technology.* Van Nostrand Reinhold, New York

Aranoff, S. (1989) *Geographic Information Systems: A Management Perspective,* WDL Publications, Ottawa

CEC (Commission of the European Community) (1990) *Directive on Public Access to Environmental Information,* EN 5222/90, Brussels.

Dale, P. F. (1991) The challenge ahead. *Proceedings of Mapping Awareness Conference,* London, pp. 333-9.

Epstein, E. F. (1987) Litigation over information: The use and misuse of maps. *Proceedings International Geographic Information Systems symposium: The Research Agenda,* Association of American Cartographers, Washington DC, Vol 1, pp. 177-84.

Epstein, E. F. (1991) Legal aspects of GIS. In: *Geographical Information Systems Volume1 Principles.* Eds. Maguire, D., Goodchild, M. F. and Rhind, D., Longman, London pp. 489-502.

Epstein, E. F. and Roitman, H. (1990) Liability for information. In: *Introductory readings in Geographic Information Systems.* Eds. Peuquet, D. J. and Marble, D. F., Taylor and Francis, London, pp. 364-71.

Rhind, D. (1992a) Data access, charging and copyright in GIS. *International Journal Geographical Information Systems,* Vol 6, No 1, pp. 13-30.

Rhind, D. (1992b) Policy on the supply and availability of Ordnance Survey information over the next five years. *Proceedings Association for Geographic Information Conference,* Birmingham, pp. 1.22.1-8.

Robbins, R. (1992) The Automobile Association's digital database of Europe. *Proceedings Mapping Awareness Conference,* London, pp. 373-382.

Smith, G. and Parr, R. (1989) *Valuation of Intellectual Property and Intangible Assets.* Wiley, New York

Wallwork, P. and Goodwin, R. (1992) The computerised Street and Road Works Register - Sharing information for improved coordination. *Proceedings Association for Geographic Information Conference,* Birmingham, pp. 2.19.1-3.

7

Standards for spatial information

7.1 The role of standards

The development of any new Information Technology leads to the adoption of standards to regulate the way the broad community of vendors and users operates. They are key elements to the process of data integration, bringing together disparate data sets (Guptill, 1991).

Some of these are achieved by general consensus and are adopted or acknowledged by a large part of the community. These become de facto standards. Alternatively standards may be developed at national or international levels and may be adopted by agreement to regulate a wider community. Such standards may be legally binding.

Many standards exist within the Information Technology industry, covering computer hardware components, communications, software, such as databases and programming languages. These have developed over many years and many continue to be revised and further developed. At the same time new standards are constantly being formulated as the industry and the user communities change.

Inevitably the development of standards lags some way behind the growth in a particular sector, and in the case of GIS with its rapid expansion in the last decade, standards are only now being developed. In this chapter the specific standards that exist within the GIS community are reviewed and the future role of standards is discussed.

7.2 Standards organisations

There are a number of organisations, both national and international, that play an important role in the development of standards that relate to spatial information. The International Standards Organisation (ISO) develops standards on a whole range of issues, but most importantly for the GIS community developed ISO 8211 which is a data description file for data exchange (see Table 7.1).

A number of national organisations have become involved in the creation of standards for spatial data, such as the US National Institute of Standards and Technology and the British Standards Institute (BSI).

In the UK the development of the National Transfer Format into a national standard (BS 7567) was undertaken by the Association for Geographic Information (AGI). Likewise in many other countries bodies associated with the main mapping agencies or umbrella organisations, both quasi-governmental or commercial, have become involved in establishing standards.

Elsewhere in Europe, the Technical Board of the Comité Européen de Normalisation (CEN) formed a Technical Committee (TC 287) to undertake the definition of standards for geographic information. This work, which commenced in 1992, will have to consider the proposals made by CERCO (French acronym for the European Committee of the Heads of Official Mapping Organizations) for a central format into which national formats would interface (Sowton, 1992a). CERCO suggested this would form the basis of a European Transfer Format (ETF) described by Sowton (1992b).

7.3 Data transfer standards

The main role of standards in GIS is to facilitate the integration of data sets from various sources. Data transfer is necessary from data capture subsystems and in exporting complete data sets to other systems. The format and structure for holding geographic information is likely to differ between computer systems so the exchange of information requires the use of standardised formats which are understood by both the originator and the recipient of the data. The format should contain all the information required to interpret it, and must be system independent.

The adoption of data transfer standards has a number of benefits to the user (AGI, unpublished):

- They allow the transfer of digital information between non-compatible systems while preserving the meaning of the data being transferred.
- Data quality information can be supplied, allowing users to evaluate the data.
- They offer the opportunity to share project costs by sharing data, resulting in lower costs for obtaining and maintaining data.
- They support efforts to update a database using multiple sources.

Data transfer requires a common understanding of how data describe the corresponding real world phenomenon, which involves the adoption of standard data models. It also requires a standard method for encoding data, creating a feature and attribute coding scheme, to set up standard classifications for various user groups.

The creation of data transfer standards usually has to take into account the diversity of existing data. This limits the ability to define a rigid data model and restricts the establishment of a well-specified feature coding scheme.

To implement a transfer format each system will require software utilities to convert data from the internal system format into the transfer format when exporting and to convert the transfer format to the system format when importing (Figure 7.1).

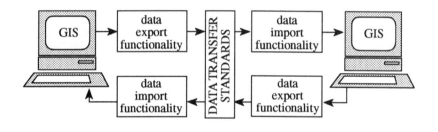

Figure 7.1 The role of data transfer standards.

The number of transfer formats that exist has risen considerably. Some are purely internal formats developed within a single organisation or group of organisations, others are intended as more broadly based standards. Table 7.1 is a lengthy summary of the main data transfer formats currently available. It is given here to show the various types of standards that are available, the inevitable plethora of national standards and to demonstrate the problems there are in establishing international standards that will be widely adopted.

The existence of so many formats will inevitably hinder the transfer of data and reduce the ability of GIS systems to integrate data from various sources. In the next few years the vast number of transfer formats will have to be refined to a dominant few. The more successful of these will inevitably form the basis of future international standards.

1. International transfer standards

- **DIGEST** Digital Geographic Information Exchange Standard (DIGEST) developed by the Digital Geographic Information Working Group (DGIWG), whose members are drawn from military survey agencies from a number of NATO countries. It is intended for bulk transfer of data and supports raster, vector (including topology) and matrix data.

- **IHO** Transfer format for digital hydrographic data published by the International Hydrographic Organisation (IHO), which is an intergovernmental organisation. The standard was published in IHO Special Publication 57 (SP57) in June 1991. It is designed for the exchange of hydrographic data to support nautical paper charts, digital charts and hydrographic surveying.

2. National transfer formats

- **ATKIS** German Authoritative Topographic Kartographic Information System (ATKIS). The standard includes a definition of data model, feature catalogue and symbol catalogue.

- **AVS** Swiss Amtliche Vermessungs Schnittstelle (AVS) transfer system for cadastral data.

- **CCSM** Canadian Council on Surveying and Mapping (CCSM) standard for data classification, feature definitions, feature codes, quality evaluation and digital topographic exchange.

- **DEM** Digital Elevation Models (DEM) format used by US Geological Survey for standard terrain data transfer.

- **DLG** Digital Line Graph (DLG) is used by the US Geological Survey for the transfer of fully structured data from the National Cartographic Database. The structure data format is DLG-3, while DLF-E is an enhanced version which a cartographic feature level.

- **GBF-DIME** Geographic Base File (GBF) is the US Bureau of Census standard for topologically structured data, first used in 1980.

Table 7.1 A summary of the main data transfer standards. Part 1.

- **GDA** Geographic Document Architecture (GDA) is being developed by the Canadians for use in an Open Systems Interconnection (OSI) environment.

- **KF85** Simple vector data transfer format developed by Swedish Association of Local Authorities.

- **MACDIF** Map and Chart Data Interface Format (MACDIF) proposed by Canadian hydrographic agencies for data transfer.

- **NES** National Exchange Standard (NES) for Digital Geographically Referenced Information is the national standard for The Republic of South Africa.

- **NTF** National Transfer Format (NTF) is the official data transfer standard for the United Kingdom. NTF was published as British Standard 7567 in 1992 (see Section 7.4)

- **ONORM A 2260** A transfer format proposed by the Austrian Standards Institute in 1990. Oriented towards large scale maps and drawings.

- **SAA** The Australian standard (AS 2482-84) for the interchange of feature coded digital mapping data.

- **SDTS** Spatial Transfer Data Standard (SDTS) is a US transfer standard published in 1991 and supporting vector, raster and relational data. Approved in 1992 as a Federal Information Processing Standard (FIPS) Publication 173 (see Section 7.5).

- **SLF** Standard Linear Format (SLF) is a US military format.

- **SPDFDM** The Standard Procedure and Data Format for Digital Mapping (SPDFDM) is Japanese specification which includes data quality, feature coding system and transfer format.

- **SOSI** Norwegian Standard for data transfer which currently only deals with vector data but with plans for raster and pictorial data.

Table 7.1 A summary of the main data transfer standards. Part 2.

- **TIGER** Topologically Integrated Geocoding and Referencing (TIGER) was developed by the US Bureau of Census for the 1990 census (see also GBF-DIME).

- **VHS** Finnish standards which deal with geographic data representation (VHS 1041) and message description (VHS 1040) and operate in an EDIFACT (ISO 9375, Electronic Data Interchange For Administration Commerce and Trade) environment.

3. National and international de jure standards

- **CGM** The Computer Graphics Metafile (CGM) standard (IS 8632) defines a picture capture mechanism and relates to other ISO standards such as GKS (Graphics Kernal System).

- **DCW** Digital Chart of the World (DCW) is a US Defense Mapping Agency vector data set based on the 1:1,000,000 scale Operational Navigational Charts (ONC). It is an implementation of the Vector Product Format (VPF) from Environmental Systems Research Institute (ESRI) and is equivalent to VRF in DIGEST (see Section 1).

- **EDIGeO** A standard devised by the Conseil National de l'Information Géographique (CNIG), similar in many respects to DIGEST. It has topological, network, spaghetti and matrix data models, defines a coding format and and has a coding catalogue for features and attributes.

- **IGES** The ANSI Initial Graphics Exchange Standard (IGES) is intended for CAD graphics. There are no facilities for handling attributes.

- **ISO 8211** ISO 8211 (BS 6690) is a specification for a data description file for information interchange. It imposes a file structure on a transfer which consists of a dictionary describing what data might occur and then the data itself.

- **STEP** Being developed by ISO to replace IGES.

Table 7.1 A summary of the main data transfer standards. Part 3.

4. Established de facto and de jure standards

- **DXF** Drawing Exchange format (DXF) was designed to allow the exchange of technical drawings between AUTOCAD and other CAD packages. It is a widely used de facto convention in the engineering and construction industries.

- **GIMMS** Geographic Information Manipulation and Mapping System (GIMMS) is a topological data transfer format used by the UK's Department of the Environment.

- **REMOTE SENSING STANDARDS** There are a number of these including Band Interlaced by Line (BIL), Band Interlaced by Pixel (BIP) and Band Sequence (BSQ).

- **SIF** Standard Interchange Format (SIF) is the Intergraph standard. It is a CAD transfer format, designed with little regard for other GIS systems.

- **SVF** Single Variable File (SVF) is an Environmental Systems Research Institute (ESRI) standard used for the transfer of raster data between GIS and remote sensing systems.

Table 7.1 A summary of the main data transfer standards. Part 4.
(AGI, unpublished)

The proposed European Standard, ETF, may be a contender, as will the US SDTF (see section 7.5).

It is not possible to consider in detail the large number of existing standards summarised in Table 7.1. However, it is important to understand the technical complexity and preparation time necessary for the development of an enduring standard that can be used widely by all or part of the geographic information community.

Some example standards are looked at in a little more detail here. The development of the UK's NTF standard and the US Spatial Data Transfer Standard are described to show the time and effort necessary to establish even a nationally accepted standard. The technical issues relating to international standards are mentioned in the context of DIGEST and finally there is brief review of a raster standard.

7.4 UK National Transfer Format

The National Transfer Format (NTF) Steering Committee was established in 1985 to provide standards suitable for the transfer of geographic and cartographic data. A wide range of user groups was represented on the original group to provide a broad-based standard that would have wide applicability.

A draft format was circulated at Auto Carto London in 1986 (Sowton and Haywood, 1986). The first draft contained a large number of feature codes which were subsequently greatly reduced after the draft consultation phase. These changes were in part due to the demands of the utility sector for a format that could result in the digital conversion of the large scale map base more quickly than hitherto planned.

Version 1.0 of the UK National Transfer Format (NTF) was first released in 1987, to be followed in 1989 by version 1.1. The responsibility for NTF was originally vested in the Ordnance Survey of Great Britain but in mid-1989 it was formally brought under the umbrella of the Association for Geographic Information (AGI). The AGI is a non-governmental organisation set up following the publication of a government report into the use of geographic information in the UK (DOE, 1987). It acts as an umbrella organisation which promotes the development of GIS.

In 1991 a draft of version 1.2 was issued for public comment by the British Standards Institution (BSI) and in June 1992 it was published as as a British Standard (BSI, 1992) as version 2.0 which incorporates both the NTF version 1.2 format and the NTF/ISO 8211 format. The accompanying documentation was completely rewritten to conform with British Standard requirements (Sowton, 1992c). The standard comprises three parts:

- Specification for NTF structure
- Specification for implementing plain NTF
- Specification for implementing NTF using BS 6690, a British Standard equivalent to ISO 8211 (see Table 7.1). This is an alternative implementation of NTF.

NTF is designed to meet the variety of solutions used by organisations using data from the national cartographic database. Simple data solutions should be made available for simple requirements and to meet this the standard has five levels, as shown in Table 7.2. As a standard NTF gives a fully specified data model in levels 1, 2 and 4 and a generic/customisable data model in levels 3 and 5. It also allows users to make use of de facto standards for raster data.

- Level 1:   Vector spaghetti. This model would only be useful where one attribute exists per feature.

- Level 2:   Vector spaghetti as Level 1 but with multiple attributes. This conforms to the DIGEST spaghetti model but has the added advantage of being able to handle cartographic text as an object.

- Level 3:   Partial topology. This level handles a variety of models including partial topology, link and node and spaghetti with closed areas. The link and node model can conform to the DIGEST chain-node model.

- Level 4:   A rigorous topological model which conforms to the DIGEST topological model. Polygon seeds are allowed.

- Level 5:   Formerly level 4, this allows a custom model to be defined based on the Level 3 data model; the use of a data dictionary is mandatory at this level. This becomes more simple to use with the introduction of ISO 8211.

Table 7.2 Five levels of the National Transfer Format (after Sowton, 1992c).

7.5 US Spatial Data Transfer Standard

A collaboration between the US Geological Survey and the US National Institute of Standards and Technology led to the development of a set of standards for earth science data and their representation. This set of standards is intended for use within Federal Government agencies but will also be available for state and local governments, the private sector, and research and academic organisations.

The development of the Spatial Data Transfer Standard (SDTS) took 10 years and in 1991 was issued by the National Institute of Standards and Technology as a proposed standard, FIPS 173. Following public consultation, an edited version of FIPS 173 was approved in July 1992, with the FIPS 173 implementation effective from February 1993. Use of the standard is mandatory for Federal agencies by February 1994 (Neff, 1992).

The objectives of the standard are (Guptill, 1991):

- To provide a mechanism for the transfer of digital spatial information between non-communicating parties using dissimilar computer systems, and reducing to a minimum the need for information external to this standard concerning the transfer.
- To provide, for the purpose of transfer, a set of clearly specified spatial objects and relationships that can represent real-world spatial entities, and to specify the ancillary information that may be necessary to accomplish the transfers required by the cartographic community.
- To provide a transfer model that will facilitate the conversion of user-oriented objects, relationships and information into the set of objects, relationships and information specified by the standard for the purposes of transfer such that their meaning will be preserved and can be discerned by the recipient of a conforming standard.
- To ensure the implementation of this standard can have the following characteristics:

  - the ability to transfer vector, raster, grid and attribute data, and ancillary information;
  - the implementation methodology can be media independent and extendable to encompass new spatial information as needed;
  - an internally contained description of the data types, format and data structures such that the information items can be identified and processed into the user's native system;
  - the data and media formats should be based where practical on existing FIPS, ANSI, ISO or other accepted standards.

The standard consists of several components:

- *Definitions and References*, including a conceptual model of spatial data and definitions of fundamental cartographic objects.
- *Transfer Specification* which defines the logical file structure for the transfer of data.
- *Data Quality* information which defines the format of the quality report. The data provider has to supply detailed information about the data set so that the receiver of the data can evaluate its fitness for particular uses.
- *Cartographic Features* which gives standard definitions for cartographic entities.

The US Geological Survey is responsible for the maintenance of FIPS 173, and also has the task of encouraging its acceptance by both users of spatial data and the vendors of spatial information systems. This it is undertaking by an extensive programme of awareness. The acceptance and the success of the standard will depend on the awareness and understanding of the role and impact of using such a standard (Neff, 1992).

7.6 DIGEST

The need for digital geographic data standards within the international defence community led to the formation of the Digital Geographic Information Working Group (DGIWG) in 1983. The remit of the group was to encourage the development of standards for both civil and military use. Although not an organisation run by the North Atlantic Treaty Organisation (NATO), the DGIWG is essentially formed of countries from the NATO alliance.

The main aim of DGIWG was to ensure that 'different national geographic information systems can exchange data effectively and efficiently with no loss of information' (Ley, 1992). This includes the exchange of data between data production agencies, between agencies and users and between users, while at the same time allowing each organisation to choose its own system specification. This requires the development of well-defined data models and consistency in the way accuracy, feature coding, geo-referencing, precision and spatial resolutions are handled.

In 1991 DGIWG published the Digital Geographic Information Exchange Standard (DIGEST), a family of internationally agreed standards which make use of existing ISO standards (DFMA, 1991).

DIGEST has six defined data models:

- Vector spaghetti or unstructured
- Vector chain/node
- Vector topological
- Raster, supporting multi-colour graphics
- Matrix or arrays of non-radiometric information, e.g. soil types or elevations. It is the same as the raster data model.
- Vector relational format (VRF) which uses faces, edges and nodes to represent vector data but linkages are supported by a number of relational tables.

The VRF model is of particular interest since the processing and computer memory requirements are less and therefore the model is better suited to a PC environment. For this reason the Digital Chart of the World product on CD-ROM, which is aimed at the PC-user, is structured in VRF.

Not only does DIGEST have data models, information about data organisation and data structure but it also includes data quality statements. At certain levels these are required as part of the standard and include such topics as source, positional and attribute accuracy, logical consistency, completeness and currency (Ley, 1992). This enables data recipients to know the quality of the data and to effect integration with other datasets.

DIGEST also contains cartographic annotations which may relate to specific or general features or locations. The important aspect of this element of the standard is that the data model and structure used are independent of the cartographic symbolisation.

The other important advantage of DIGEST is that it provides a standard method for coding features and attributes. This is given in the Feature and Attribute Coding Catalogue (FACC), which is a dynamic document whose changes are closely controlled. The benefit of this system is that it is not necessary to provide a comprehensive data dictionary with each data transfer. The many-to-many type data transfers are far more efficient as they only require one look-up table (Ley,1992).

The development of DIGEST has also taken into account other national exchange standards. For example, there are compatible profiles between DIGEST and the latest versions of the UK National Transfer Format (Rowley and Sowton, 1991). It should also be possible to exchange data between standards without loss of information.

7.7 ASRP Raster Standard

A particular aspect of the DGIWG work has been the development of a raster data standard, which is particularly important for graphic display in military applications. The result is the ARC Standard Raster Product (ASRP) which was published in 1991 (DGMS, 1991).

ASRP provides a reference system that enables a global seamless dataset to be achieved without necessary transformations, sizing or cutting and patching that would be required were the data retained in map sheets on their original datum and projection (Gennery, 1992). An ASRP dataset includes header information about the contents of the dataset, source information about the base graphics, quality information on accuracy and quality of the image and the pixel data.

ASRP data uses WGS84 (the World Geodetic System established in 1984) as a common reference system and the equal ARC-Second Raster Chart/Map (ARC) as the projection and co-ordinate system. The ARC system divides the Earth's ellipsoid (the mathematically computed surface) into 18 latitudinal bands or zones. The transformation into ARC zones involves some distortion due to the projections used but it is considered to be indiscernible on a computer screen (Gennery, 1992).

The standard uses the Extended Colour Coding (ECC) technique which aims to maximise the data on colour while using the least volume of data. This uses one colour code for each colour on the graphic but also introduces transition codes, for example tones of pink between red and white, which may be incorporated to aid definition of text, fine lines and areas of high contrast. This lacks some flexibility over RGB data but is better for post processing.

The use of transition colours is also intended to help remove some of the inherent problems that arise from map scanning (Gennery, 1992):

- Anti-aliasing: Use of transition codes between primary colours to reduce high contrast and the serrated effect of pixellation of a smooth line.
- Dot patterns: Printing screens and dot hachures are more easily retained as recognisable patterns when supported by transition colours.
- Image rescaling: Transition colours provide a higher level of information on which to base resampling techniques.

The pixel resolution used by ASRP is 100 dots per centimetre. At a scale of 1:50,000 a distance of 0.5 km is represented by a nominal 100 pixels. ASRP data are provided in a run-length encoded form.

7.8 Other standards

While this chapter has focused on the data transfer standards which are so critical to the development of the spatial information handling community, there are other standards that are also important.

Vendors of GIS software or turnkey solutions of hardware, software and occasionally data, such as that offered as part of a collaborative project in the UK by Digital, Laserscan and the Ordnance Survey in 1992, have to consider existing standards in the computer industry. These include de jura and de facto standards for nearly all aspects of computing from microchips to cabling and connectors.

There are standards related to the development of storage media, for example ISO 1001 for magnetic tapes (ISO, 1986) and ISO 9660 on the volume and file structure of CD-ROMs (ISO, 1988), as well as standards on open systems (ISO, 1987). For a more detailed review of standards that pertain to the computer industry readers should consult the general computing literature.

There are other related developments which are also important, with

- Level 0: The specification, which defines what the Land and Property Gazetteer contains in terms of data content, data quality and standards used. This should be application independent.

- Level 1: The Gazetteer built-up directly from the data specification. As a minimum this will include some compulsory fields.

- Level 2: Extended Gazetteer will included extra data added to each record, either as optional fields of new fields. Products of this type will be user specific such as a link to a system of potentially contaminated land.

- Level 3: Integrated applications where the gazetteer is linked to larger scale existing or yet to be created systems.

Table 7.3 Proposed structure of the UK Land and Property Gazetteer standard (after Pugh and Cushnie, 1992).

the introduction of standards for non-graphic spatial data types, an area where there are currently few standards. An example is the progress towards a national standard for a land and property gazetteer in the UK (Pugh and Cushnie, 1992), see Table 7.3. This will enable local authorities and other users of such data to transfer their property gazetteers, which have existed for many years, much more easily.

This development will also support the proposed UK National Land Information System, Domesday 2000 (Dale, 1992). Standards for the exchange of information to meet the needs of a national LIS have to cover not only transfer formats but also data classification and accuracy standards (Dale and McLaughlin, 1988).There is also the need to evaluate the quality and compatibility of sources, particularly where they relate to the compilation and continued maintenance of a cadastral base.

7.9 Importance of standards

Standards are important in achieving effective integration for spatial data. The development of comprehensive and widely acceptable standards requires considerable work; DIGEST took over 30 man/years of effort to reach the publication stage (Ley, 1992). They also needs extensive co-operation and financial support amongst interested groups. Standards are expensive to produce and unless funded by government or the relevant industry, or both together, are unlikely to be developed in a comprehensive fashion and will therefore gain little support.

When established data transfer standards can underpin the development of integrated information management between data generators, system suppliers and the wide variety of users from many sections of the broad spatial data community. To this end the national and international map-based standards are critical to the display and manipulation of map-based information.

Standards are necessary to underpin integration not only for hardware and software supplied by the GIS vendor community but also for the data suppliers and data users, whether that data is graphic or non-graphic (Figure 7.2). Rowley (1992) indicated that standards are the single most critical factor in the long term success of the geographic information market.

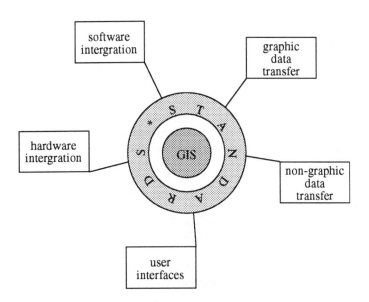

Integrated Information Management Systems

Figure 7.2 The importance of standards in developing integrated GIS solutions.

Integrated solutions and the development of desktop approaches will require a closer conformance to standards if the users are to be able to upgrade their GIS solutions as new packages become available, transfer data without the expense of data conversion and be able to use or pass on 'value added data'. In addition, standards that bridge the gap between GIS and other information management systems, such as those for document archiving or image management, will also need to be included in broader-based strategies for spatial data standards.

References

AGI (Unpublished) Data transfer formats - A review. Report by the Association for Geographic Information Standards Committee.

BSI (1992) *BS 7567: 1992 Electronic Transfer of Geographic Information (NTF), Part 1 Specification for NTF structures*, London.

Dale, P. F. and McLaughlin, J. D. (1988) *Land Information Management.* Clarendon, Oxford.

DFMA (1991) *DIGEST Edition 1.0,* Defense Mapping Agency (on behalf of DGWIG), Washington D.C.

DGMS (1991) *The Digital Geographic Information Working Group International Specification for ARC Standard Raster Product.* Director General Military Survey, Feltham, Middlesex.

DOE (1987) *Handling Geographic Information.* Report of the Committee of Enquiry chaired by Lord Chorley, HMSO, London.

Gennery, D. (1992) A working raster standard - ASRP (ARC Standard Raster Product). In: *Geographic Information 1992/93. The Yearbook of the Association for Geographic Information.* Eds Cadoux-Hudson, J. and Heywood, D. I., Taylor and Francis, London, pp. 398-403.

Guptill, S. C. (1991) Spatial data exchange and standardization. In: *Geographical Information Systems: Principles and Applications.* Eds. Maguire, D., Goodchild, M. F. and Rhind, D. W., Longman, London pp. 515-30.

ISO (1986) *ISO 1001 Information Processing - Files structure and labelling of magnetic tapes for information interchange.* Edition 2.

ISO (1987) *ISO 8824 Information Processing Systems - Open Systems Interconnection - Specifications and Abstract Syntax Notation One.*

ISO (1988) *ISO 9660 Information Processing - Volume and File Structure of CD-ROM for Information Exchange.* Edition 1.

Ley, R. (1992) The Digital Geographic Information Exchange Standard - DIGEST. In: *Geographic Information 1992/93. The Yearbook of the Association for Geographic Information.* Eds. Cadoux-Hudson, J. and Heywood, D. I., Taylor and Francis, London, pp. 392-7.

Neff, K. (1992) The spatial data transfer standard (FIPS 173): A management overview. *Proceedings GIS/LIS,* San Jose, pp. 614-7.

Pugh, D. and Cushnie, J. (1992) The land and property gazetteer. *Proceedings Association for Geographic Information Conference,* Birmingham, pp. 2.20.1-6.

Rowley, J. (1992) A strategy for geographic information standards. *Proceedings Association for Geographic Information Conference,* Birmingham, pp. 2.21.1-7

Rowley, J. and Sowton, M. (1991) National Transfer Format - Progress Report. *Mapping Awareness,* Vol 5, No 5, pp. 25-7.

Sowton, M. (1992a) British Standard BS 7567 - Electronic transfer of geographic information (NTF). In: *Geographic Information 1992/93. The Yearbook of the Association for Geographic Information.* Eds. Cadoux-Hudson, J. and Heywood, D. I., Taylor and Francis, London, pp. 387-91.

Sowton, M. (1992b) From National to Normalised Transfer Format. *Proceedings DRIVE Workshop - European Digital Road Map.*

Sowton, M. (1992c) The National Transfer Format - the path towards a British Standard. *Proceedings Mapping Awareness,* London, pp. 321-31.

Sowton, M. and Haywood, P. E. (1986) National standards for the transfer of digital map data. *Proceedings Auto Carto London.* Ed. Blakemore M., pp. 298-311.

8

Analysis functions in GIS

8.1 Analysis in GIS

The range of analysis functionality in existing GIS software which may be used to query and manipulate spatial information is extensive. Less well developed are analytical tools which utilise modelling and statistical methods.

It is not the purpose of this book to review GIS analysis functions in any detail; these may be found in many texts, for example Burroughs (1986), Aranoff (1989) and Laurini and Thompson (1992). Aranoff (1989) produced a classification of the main GIS analysis functions, reproduced here as Table 8.1. However, it is necessary to look at the development of analytical functions within spatial information systems and to consider their role in the decision-making process.

A key issue in the design and implementation of a spatial information system is the type of analytical capabilities required to achieve the system objectives. Many current applications which are grouped under the generic title of GIS are essentially digital mapping or CAD solutions, with a limited number of analytical functions, usually based on spatial query.

A measure of the growing maturity of the GIS software market is the increased range of query and manipulation functionality available with the latest version of many systems. The selection of a GIS must therefore be based not only on the range of functionality but also on the efficiency, accuracy and quality of the analyses and how the results from analytical processes are managed.

8.2 Manipulation versus analysis

What distinguishes GIS from other information systems is the ability to manage and analyse spatial data. The general classification of analysis functions given in Table 8.1 simplifies the range of analyses found in most GIS packages into a number of generic groups. However, all the functionality in this list should more properly be regarded as data

Maintenance and analysis of the spatial data	Format transformations
	Geometric transformations
	Transformations between map
	    projections
	Registration of detail from
	    different sources
	Edge matching
	Editing of graphic elements
	Generalisation	

Maintenance and analysis of the attribute data	Attribute editing functions Attribute query functions	
Integrated analysis of spatial and attribute data	Retrieval/classification/ measurement	Retrieval Classification Measurement
	Overlay operations	
	Neighbourhood operations	Search Line-in-polygon and point-in-polygon Topographic functions Interpolation Contour generation
	Connectivity functions	Contiguity measures Proximity Network Spread Seek Intervisibility Illumination Perspective view
Output formatting	Map annotation Text labels Texture patterns and line styles Graphic symbols	

Table 8.1 A classification of GIS analysis functions (Aranoff, 1989).

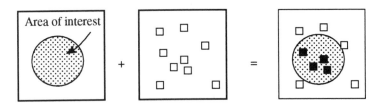

1. Proximity analysis where a subset of data is extracted on the basis of a user defined area

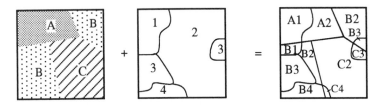

2. Overlay operation to create a new set of spatial objects whose attributes are a composite of the derivative data sets

Figure 8.1 Examples of manipulation processes.

manipulation rather than analysis. Analysis implies quantitative, usually statistical, methods for interpreting data (Openshaw, 1991). Most GIS users refer to analysis when they mean manipulation. The manipulation functionality is usually adequate since it provides the ability to select and reselect subsets of data for display and visual interpretation or to generate tabulated summaries.

Many users of GIS do not have the skills to make use of more sophisticated tools, such as spatial statistics or operation analysis techniques. One of the limiting factors in the wider use of remote sensing has been the skills required to carry out image processing operations. The greater use of spatial information systems implies that the majority of users will have only a limited understanding of the data issues, including what analytical functions may be appropriate to their problem and how to interpret the results from statistical analyses.

Very few existing GIS systems have any truly analytical functionality (Goodchild, 1991). A summary of GIS, digital mapping, remote sensing and facilities management systems is published annually (Eynon, 1992) in which system functionality is detailed. The analysis functionality listed in this summary is predominantly concerned with data query and manipulation.

The 1990s have seen GIS begin to embrace statistical techniques, mathematical modelling and operational analysis as part of a more comprehensive analytical toolbox. Such capabilities are generally limited to research environments and specialised application areas where particular functionality is added 'in-house' to GIS packages or comes from other information systems. As an example, a highways department may use a GIS to identify the geographical location of a road section, to which may be added details about surfacing and structure, but the models used to determine when the road needs to be resurfaced are typically accessed from engineering systems.

8.3 Query and manipulation in integrated solutions

The majority of GIS packages have capabilities for interrogating data which can be categorised as either query functions or manipulation functions. Query functions allow the user to search the spatial data and its associated databases for particular objects or groups of objects based on their object classification or attribute information. It is possible to select all the line segments that constitute the generic group road or to identify those segments that constitute a particular road based on its network number.

The manipulation functionality involves the sampling or resampling of one or more spatial data sets to create subsets of existing entities or to create new spatial entities. It is possible to identify all the features in a generic class that exist within a certain distance of a defined location or to combine datasets to create new objects (Figure 8.1).

The value of query and manipulation functions will depend on the problems which are being addressed. Together they have proved very powerful, particularly when a range of functions is used in a prescribed sequence to solve a problem. Figure 8.2 is an example of how four simple data sets may be interpreted using a number of manipulation processes to identify areas suitable for afforestation.

The ability of a system to carry out manipulation functions will vary and, while most GIS packages will contain a broad range of such functionality, the efficiency with which they carry them out will depend on a range of factors, such as data structure used, algorithms and original design objectives.

In an integrated environment the management of the various data sets produced during query and manipulation processes is important.

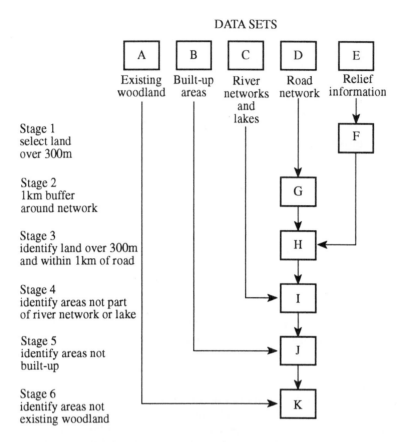

Problem: To identify areas suitable for afforestation that are over 300 metres above mean sea level, within 1 km of an existing road and not existing woodland, built up area or a river or lake.

Figure 8.2 An example of a structured approach to solving a problem using overlay and buffering processes.

The original data will need to be accessed by the appropriate users for query and manipulation processes but update authority must be strictly controlled. The results of a manipulation process must not be able to over write the original data. In some environments copies of the original data will be made to allow the user to carry out the manipulation, but this can be costly in both storage and processing time.

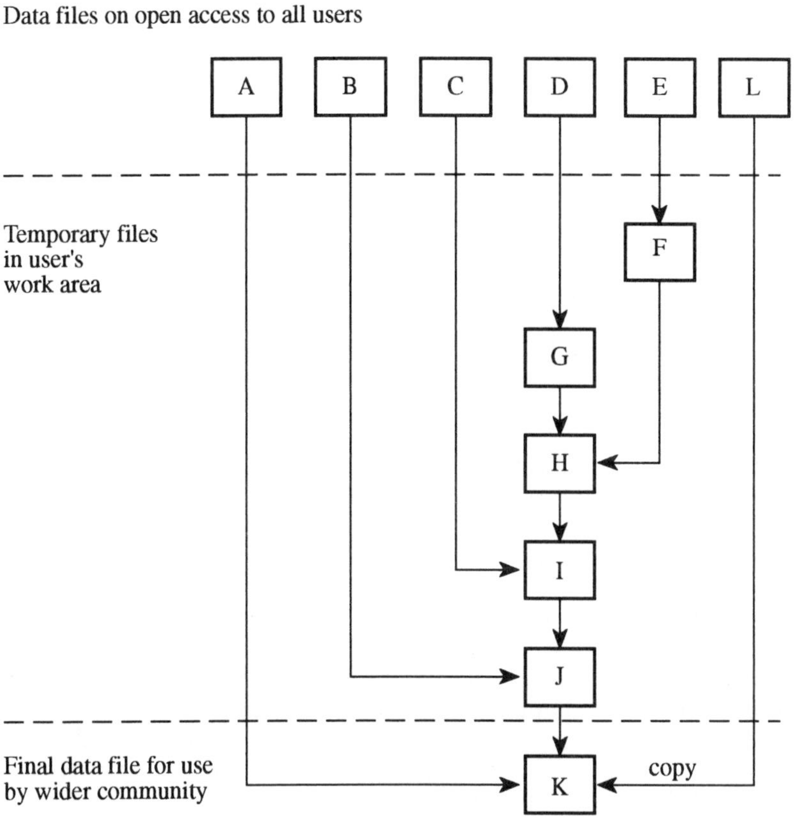

Data files on open access to all users

Figure 8.3 User access to data sets created in a manipulation process.

In most systems the results of a manipulation are written to the user's workspace as some form of temporary file. These may be retained by the user until such times as the manipulation is complete or the files are superseded by latter versions. Such temporary versions of the data sets would usually only be accessible to the individual who created them and those with system manager access, and not the wider GIS user community.

The final result of the manipulation may be retained for use by the individual, with no general access to the results, or may be made widely

- Pattern spotters and testers
- Relationship seekers and provers
- Data simplifiers
- Edge detectors
- Automatic spatial response modellers
- Fuzzy pattern analysis
- Visualisation enhancers
- Spatial video analysis

Table 8.2 Basic generic spatial analysis procedures (Openshaw, 1991)

available by storing the result in an open access location. In Figure 8.3 the datasets derived from the analysis in Figure 8.2 are shown in terms of their user access.

An important component in this process of creating new data sets which have varying levels of user access is the need to maintain comprehensive records. These need to cover aspects such as the data sources and the currency and quality of the derived data. In this respect metadata systems are very important in the successful control of data sets derived from manipulation processes (Blakemore, 1991).

Without these it is possible to lose control over the type and range of products generated. This is important from the overall system level down to the individual user level (see Section 8.7)

8.4 Statistical analysis

It has been argued that the introduction of quantitative techniques into spatial information systems would considerably enhance the end-user's analysis options (Openshaw *et al.*, 1991). So far there has been little progress towards incorporating the range of existing techniques into current products with the result that most systems have comparatively simple analytical functions (Goodchild, 1991).

One solution is to link GIS with existing statistical packages such as SAS, SPSSX or Minitab. The problem with this approach is that the statistical packages do not provide the analytical functionality needed for GIS since they were not developed to handle spatial data. Such linked solutions would require the development of modules to ensure that spatial data structures are not lost. This would be difficult to achieve

Type of geographical data	Methods of analysis
Point	Nearest neighbour Quadrat methods
Line	Network analysis and graph theoretic methods Fractal dimension Edge detection
Area	Shape measures Spatial autocorrelation Spatial regression Regionalization Spatial interaction Location-allocation modelling
Surface	Image processing Bayesian mapping

Table 8.3   A simple typology of some spatial analysis methods (Openshaw, 1991).

given the lack of standards in data structures across the current range of GIS solutions.

A second approach is to add statistical functionality directly into GIS packages by the development of integrated modules. Openshaw (1991) identified a list of basic generic spatial analysis procedures that should form part of a spatial analysis toolkit for GIS (Table 8.2).

Table 8.3 is a simple typology of some spatial analysis methods (from Openshaw, 1991). It is based on the four types of geographical entity to be found in the map, although it should be noted that each of these can be mapped onto each other. For example, point data can be aggregated into areas, and data for areas can be aggregated into surfaces.

The development of these procedures is a matter of considerable interest in the GIS research community. Both the US National Center

- Response modelling for large data sets with mixed scales and measurement levels.
- Practical methods for cross area estimation.
- Zone design and spatial configuration engineering.
- Exploratory geographical analysis technology.
- Application of Bayesian methods.
- Application of artificial neural nets to spatial pattern detection.

Table 8.4 Six key spatial analytical research topics (Openshaw, 1991).

for Geographic Information and Analysis (NCGIA) and the UK's Regional Research Laboratory initiative identified a number of key analytical research topics which should be pursued through the 1990s (NCGIA, 1989; Openshaw, 1990; Goodchild *et al.*, 1992). These are summarised in Table 8.4.

The number and range of statistical techniques that exist for use with spatial data are limited. As Goodchild *et al.* (1992) note, it is only in the last few years that a realignment of spatial data analysis with other explicitly spatial data technologies such as cartography, GIS and remote sensing has finally begun. However, much of the statistical spatial data analysis is still strongly linked to a spatial data handling packages like Minitab.

There has been research in a number of areas and some of these have been developed into fairly widely used applications. An example is the technique called kriging developed by mining geologists. Kriging is a modelling technique using spatial autocorrelation, which is a numerical measurement based on the premise that places closer together in space tend to be more similar than different. This has been used in many applications such as the mapping of heavy metal pollutants on floodplains (Leenaers *et al.*, 1989).

Another example is the use of Monte Carlo Simulation for investigating different ways of aggregating spatial units, often referred to as the Modifiable Areal Unit problem (Beneditti and Palma, 1991). Fisher (1991) used Monte Carlo Simulation to evaluate the accuracy of agricultural land valuation using land use and soil information.

There is a growing number of spatial data users who have an interest in the development of spatial analysis tools. But rigorous statistical techniques are still something of an anathema to most users of spatial data whose interest in GIS stems in no small part from the visual impact of maps and images on a computer screen. This is likely to continue unless the analysis techniques that are developed can be used without a background in statistical techniques. This approach increases the problems of interpretability of results which may well be based on data that is really only suitable for low level descriptive functions.

There have been some developments towards an interactive statistical graphics system linked to a GIS. Haslett *et al.* (1990), for example, describe some prototype software which illustrates the possibilities of using statistical graphics as further views of the data, which can be made active and thus provide alternative means of querying the data. For many users this may be an adequate level of statistical functionality.

While Openshaw (1991) argues that the lack of development in spatial analysis techniques is likely to be the major impediment to the full exploitation of GIS, he none-the-less notes that there is a need to place the emphasis on insight and creativeness, such that spatial analysis for GIS may be an art and not a science. This may point the way to two different levels of GIS capability, the manipulation and display systems currently available and more specialised solutions for those with detailed analytical requirements.

8.5 Modelling of spatial data

An important aspect of introducing true spatial analysis into GIS is the development of suitable modelling techniques. This involves approaches to problem definition, the setting of aims and objectives, defining analytical processes and the evaluation of results. Modelling is typically an iterative process which should form part of a Spatial Decision Support system (SDSS).

Decision Support Systems are either data-oriented or model-oriented but increasingly they provide flexible control over models (Silver, 1991). Modelling approaches to DSS include accounting models which calculate consequences of planned actions, optimisational models which provide guidelines for action by generating the optimal solution consistent with a series of constraints, and suggestion models which lead to a specific suggested decision for a fairly structured task.

The application of modelling to spatial information in an integrated multi-user environment requires the development of well defined modelling strategies that provide feedback mechanisms both within the modelling process and to the decision makers at various levels in the hierarchy. Typically modelling will use a range of query, manipulation and true analysis techniques and can involve complex procedures that are the domain of the specialist.

Few modelling applications in GIS are suited to the user with limited statistical awareness. An example of how complex the modelling process can be is that described by Aspinall (1992). In this a modelling procedure is developed which predicts the distribution of red deer in specific areas and which is then used to predict the distribution throughout a region. The aim of the modelling procedure is to predict distributions within one data set by combining a number of other data sets. The data set combination is carried out using Bayes' theorem.

8.6 Models for managing derived data

In Chapter 3 some of the issues relating to data models were discussed. The results from analytical processes in a GIS do not affect the physical data models since derived data will be stored in a system-compatible data structure. There is, however, a need to develop management models for handling data generated from analytical processes. This has been touched upon in section 8.3, in which the various results of manipulation processes had differing levels of user access. The models for managing such data have to consider the following:

- Overall system security;
- Integrity of the original data;
- Size of the original databases and the computing costs of generating many subsets of the data, albeit for temporary use;
- Security of data that are being used to perform a particular manipulation, analysis or modelling process;
- Availability of the latest data and user awareness of updated information;
- Metadata records on currency, quality and source of individual datasets for the purposes of providing information from which to compile metadata for derived data;
- Maintenance of records on derived datasets;
- A hierarchical information structure in which there may be various levels at which data are read or manipulated.

It is necessary to develop a structure for managing such derived spatial data which takes account of the organisational structures and the computer systems. A particular problem occurs with the typical GIS which has a spatial information database interfaced to a DBMS. The results from analytical processes may result both in the creation of new spatial objects and in new attributes to existing objects or in just one or other of these. There have to be upgrade and access pathways defined to

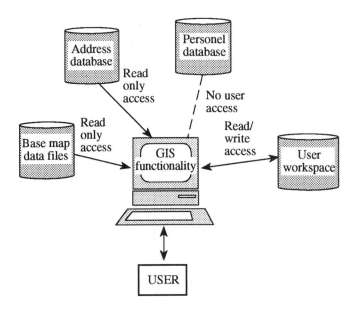

Figure 8.4 Varying levels of user access to information in a GIS.

the various database components for each individual or user group in the organisation. Figure 8.4 shows how this might apply to an individual, with read only access, write access or delete access for various database elements.

This may be developed to show how various individuals within an organisation have differing access to the information depending on their level in the hierarchy and the tasks they are required to perform (Figure 8.5). From this a three-dimensional access net may be derived based on organisational structure, computer system and information management tasks.

8.7 Metadatabases and data dictionaries

The problem of organising both the original data and any derived data in a GIS is one which is of considerable importance to the users and system manager. Every user needs to know what data are currently available, where they are located and what are the current other uses of

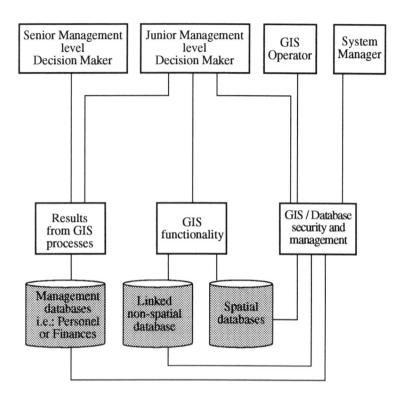

Figure 8.5 Varying levels of user access relating to organisational structure and tasks.

these data. This allows the user to judge whether or not their manipulation or analysis of the data has implications for other active user groups and to decide on the processes that should be carried out with regard to informing other users that they are to update a dataset or produce a new derivative dataset.

As an example, the planning application database of a local authority planning department may be updated once a week. Those responsible for considering current planning applications and their implications on the local environment will require the latest information to ensure accurate and consistent judgements may be made. The revision process must ensure that those who wish to review the database know when updates will occur.

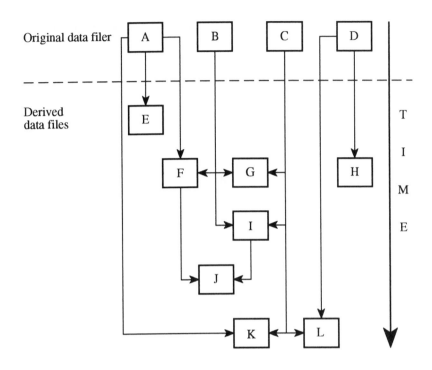

Figure 8.6 Derived data files, each of which has a different antecedence and currency.

Further problems occur when more than one person wants the same data file to derive different datasets from it. This leads to a hierarchy of derived data files, each of which has its own accuracy, quality and currency criteria. This is complicated when multiple data sets are used, as shown in Figures 8.6.

The current approach to solving this type of problem is to use a data dictionary which records the occurrence and use of every variable and program module throughout the lifetime of a system (Newman, 1991). These were originally developed for programming and subsequently used in DBMS applications. The complexity of the analytical processes in GIS means that they are not necessarily well suited for spatial information handling, especially in an integrated database environment.

The main limitation on the development of suitable metadata systems is the conflict between speed, cost and consistency (Newman, 1991). To achieve a low cost but fast collection of the necessary metadata it is usually necessary to operate at a 'local' level.

There also has to be a general perception that metadata are important and that each user is generating the necessary metadata for their derived data files. This requires rigorous operating procedures for the information system and the people who work within it and quality assurance checks to ensure consistent application of these procedures. Rigorous operating procedures often take some considerable time to establish and implement. Approaches to such aspects as metadata collection or information security and backup are often poorly developed during the GIS implementation process, leading to a poorer quality implementation and some loss of the system benefits.

8.8 Developing operating procedures for analytical processes

Most organisations have complex decision-making structures built in to their organisational hierarchy. In order to service the sophisticated information flow requirements patterns are established, some intentially, others grow out of necessity. Computer-based systems are designed to match the needs of the organisation and its decision making processes.

Where spatial data is being manipulated and analysed for the purpose of deriving information that can be used for decision making, suitable operating procedures have to be developed, often referred to as standard operating procedures (SOP). These detail the way a system should work, what the inputs and outputs are and what action is required of each element in the system at any given point in the process. The establishment of such procedures is important if the results are to be consistent. In the example given in Figure 8.2 five different data sources were used to derive a final map but in the process five other products were generated. While some steps could be undertaken in a different order to that given, the process of calculating the buffer had to be done before dataset D was added to the relief information in order to achieve the correct result.

Manipulation and analysis in a spatial framework generates large volumes of data and many intermediate data sets that will often be retained on a temporary basis while a particular process is being carried out. Procedures have to be set up that maintain a check on what spatial datasets are being generated, what their antecedence is, what they are being used and who has access to them.

Without suitably defined procedures, problems with datasets and their use are difficult to correct. Error tracking can be very difficult and systems are likely to perpetuate or even exacerbate problems. Also the lack of such procedures means that it is very difficult to adjust a system to a change in a procedure and to predict the consequences of that change.

References

Aranoff, S. (1989) *Geographic Information Systems: A Management Perspective.* WDL Publications, Ottawa.

Aspinall, R. (1992) An inductive modelling procedure based on Bayes' theorem for analysis of pattern in spatial data. *International Journal of Geographical Information Systems,* Vol 6, No 2, pp. 105-21.

Benedetti, R. and Palma, D. (1991) On the modifiable unit problem: A Monte Carlo approach. *Proceedings European conference on Geographical Information Systems,* Brussels, pp. 85-94.

Blakemore, M. (1991) Access and security issues in the provision of geographic information. In: *A Symposium on 'Approaches to the Handling of Spatial Metadata'.* Association for Geographic Information, London, pp. 26-30.

Burroughs, P. A. (1986) *Principles of Geographical Information Systems for Land Resources Assessments.* Clarendon Press, Oxford.

Eynon, D. Ed. (1992) *1993 International GIS Sourcebook.* GIS World Inc., Fort Collins.

Fisher, P. F. (1991) Modelling soil map-unit inclusions by Monte Carlo simulation. *International Journal of Geographical Information Systems,* Vol 5, No 2, pp. 193-208.

Goodchild, M. F. (1991) Progress on the GIS research agenda. *Proceedings European conference on Geographical Information Systems,* Brussels, pp. 342-50.

Goodchild, M., Haining, R., Wise, S. and 12 others (1992) Integrating GIS and spatial data analysis: problems and possibilities. *International Journal of Geographical Information Systems,* Vol 6, No 5, pp. 407-24.

Haslett, J., Wills, G. and Unwin, A. (1990) SPIDER - an interactive statistical tool for the analysis of spatially distributed data. *International Journal of Geographical Information Systems,* Vol 4, No 3, pp. 285-96.

Laurini, R. and Thompson, D. (1992) *Fundamentals of Spatial Information Systems.* Academic Press, London.

Leenaers, H., Burrough, P. A. and Okx, J. P. (1989) Efficient mapping of heavy metal pollution on floodplains by co-kriging from elevation data. In: *Three Dimensional Applications in Geographic*

Information Systems Ed. Raper, J., Taylor and Francis, London, pp. 37-50.

NCGIA (1989) The research plan of the NCGIA. *International Journal of Geographical Information Systems,* Vol 3, No 2, pp. 117-36.

Newman, I. A. (1991) Building a data dictionary: Metadatabases, data dictionaries and information resource dictionary systems. In: *A Symposium on 'Approaches to the handling of spatial metadata',* Association for Geographic Information, London, pp. 23-5.

Openshaw, S. (1990) Towards a spatial analysis research strategy for the Regional Research Laboratory initiative. In: *Geographical Information Management: Methodology and applications.* Eds. Masser, I. and Blakemore, M., Longman, London, pp. 18-37.

Openshaw, S. (1991) Developing appropriate spatial analysis methods for GIS. In: *Geographical Information Systems Volume 1: Principles* Eds. Maguire, D. J., Goodchild, M. F. and Rhind, D., Longman, London, pp. 389-402.

Openshaw, S., Brunsden, C. and Charlton, M. (1991) A spatial analysis toolkit for GIS. *Proceedings European Conference on Geographical Information Systems,* Brussels, pp. 788-96.

Silver, M. S. (1991) *Systems that Support Decision Makers. Description and Analysis.* Wiley, New York.

9

Graphical user interfaces

9.1 Human–computer interface

The human–computer interface is a particularly important component of a spatial information system. If the user-model is based on GIS system specialists passing results of analysis to a decision maker then a very different kind of interface is necessary from the model where relatively inexperienced GIS users access the data as part of an integrated information strategy for answering complex problems.

Most GIS packages are designed as generic tools in which access to the system functionality is a priority. A high level of expertise is required for the user to be able to progress through the logical sequence of commands. Improving the user interface may make the operation of the system easier but does not necessarily improve the user's understanding of whether the GIS is solving the problem in an optimum fashion (Albaredes, 1992).

The enhancement of the human–computer interface is a critical area to the development of integrated solutions. This interface has to address a number of issues:

- *The type of user and their experience of particular applications.* Users of GIS software may vary from very experienced to recently trained. Very experienced system users often prefer command driven systems since they are able to quickly and efficiently execute a series of commands with which they are familiar and understand the results that will be achieved. Less experienced users may well need prompts and advise to determine the right logical sequence of commands and their various parameters.
- *The nature of the problem and the approaches that may be adopted for its solution.* In many complex GIS systems, there are various ways of solving a particular problem, each of which may incur accuracy or efficiency penalties. The interface should be able to provide the user with the information necessary to

select an appropriate strategy. As an example, the topological structuring of data in ARC/INFO can be achieved using CLEAN or BUILD commands, each of which has a subtly different usage.

- *The need for different levels of interaction.* Some functionality may be used frequently and will be well understood. Other functionality will be little used and require a higher level of support to use. For example, a user may experience problems in finding a little used command in a complex hierarchical menu structure.
- *The need for interactive help facilities.* The interface should provide levels of user help, both to the operating system and to specific applications.
- *Provision of alternative interface strategies.* For example it may be efficient to use a mouse to call up a command from a pull-down menu when used infrequently, but it may be much more efficient to use a single key stroke to effect the same command if used constantly.
- *Issues related to the visual impact of the interface and its usability.* The importance of a visually acceptable interface to enhance the efficiency of the user can not be underestimated. A common problem is the adoption of an icon-based system in which the individual icons are not intuitive and there are often too many icons for the size of screens, leading to clutter and difficulty in interpretation.

With the advent of window-based systems productivity gains are achieved by the user being able to run multiple applications concurrently on one system. The greatest benefit is obtained where the various applications have the same interface, can communicate and share data.

9.2 User-friendly GIS

Albaredes (1992) states that 'traditional technology-driven GIS are clearly showing limitations in terms of usability'. Much exists in the literature on the implementation of GIS and application trends but there is far less on the inherent problem of GIS usability.

The user community in an integrated geo-information handling environment will range from the GIS specialist to the casual user, and yet few systems are designed for such a broad spectrum of user experience. Users with virtually no knowledge in GIS or related disciplines would find it very difficult to use most existing applications. There is still a need to develop a general awareness of spatial concepts

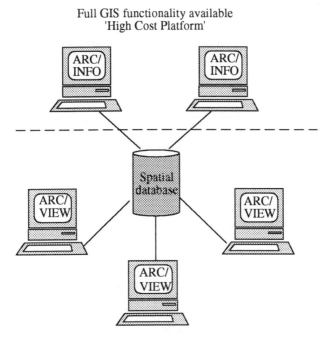

Full GIS functionality available
'High Cost Platform'

'Low cost platforms'

Data query and spatial manipulation functions only available

Figure 9.1 Distributed information model based on the use of 'viewing GIS'.

and the underlying issues relating to spatial data throughout the user community, but if these exist then using a GIS should be much more intuitive. The most important aspect of achieving this is the development of suitable user interfaces.

The evolution towards more user-friendly GIS has begun with the development of products that use data from more complex systems but offer the casual user the ability to perform queries and simple spatial manipulation with only a limited knowledge of the underlying technology. In particular products such as ARCVIEW from Environmental Systems Research Institute (ESRI), SPANS MAP from

TYDAC and the MGE Project Viewer from Intergraph Corporation are examples of this development.

These may be termed 'Viewing GIS' and allow data to be distributed after processing in a fully functional GIS to low cost platforms running cheaper applications where the data may be viewed (Figure 9.1). To be successful this model of data integration relies on the development of interfaces that allow users with little background in the core system to view the data.

A further progression of this concept is the development of GIS solutions that meet the needs of particular user communities. This may be referred to as User-Oriented GIS (Figure 9.2) and requires system vendors to develop applications related to specific sectors. Albaredes (1992) refers to this as solution engineering.

9.3 Generic interface types

Early GIS solutions had command driven interfaces and required considerable expertise to facilitate their use. These were followed by menu-driven interfaces which were operated either from the keyboard or from a mouse. Subsequently there has been the development of specialised Graphical User Interfaces (GUIs) for GIS solutions based on the window–icon–mouse–pop-up menu (WIMP) model. The use of windows and icons is intended to improve the 'touch and feel' of the software, making it more intuitive to use.

Early work on WIMP interfaces developed from research on the Smalltalk-80 project at the XeroxPARC in California in the early 1980s (Goldberg and Robson, 1983). These developments were particularly important for graphics-oriented applications such as GIS. The problem of poor interfaces was recognised as being a significant impediment to the development of spatial information handling in the UK in the late 1980s (HMSO, 1987).

Following on from the development of the WIMP interface there has been much research into effective screen design, especially in the representation of spatial analysis results (Egenhofer and Frank, 1988) and in spatial language design (Mark and Frank, 1989). Most systems now use a 'point and click' model to drive much of the interface. A mouse and screen cursor is used to replace the majority of keyboard work, thereby reducing the need for efficient keyboard skills. There are, however, still some systems which are based on command-line or fixed menu interfaces. The most effective solutions often combine elements of the various interface types, and provide alternative strategies, such as a keystroke which may be used instead of a pull-down menu item.

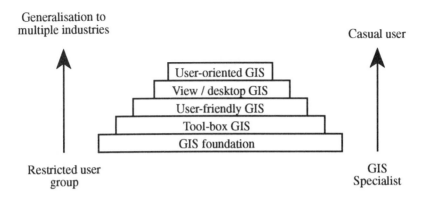

Generalisation to
multiple industries

Casual user

User-oriented GIS
View / desktop GIS
User-friendly GIS
Tool-box GIS
GIS foundation

Restricted user
group

GIS
Specialist

Figure 9.2 The evolution towards usable GIS (after Albaredes, 1992)

The following is a list of the generic types of interface functionality that are typically used:

- Command line where the user has to know and type in each command and its parameters, or at least be prompted by the system for the necessary elements.
- Fixed screen menus, from which the user may select operations by the use of keystrokes or a mouse.
- Pull-down menus which display groups of operations under generic headings.
- Pop-up menus that appear alongside a command, often giving the range of choices for a particular parameter.
- Icons which replace words with symbols or graphics that are intuitive and require no further textual explanation.
- Sound prompts for either keystrokes, mouse operations or to warn the user that some process is taking place.
- Windows that allow multiple datasets to be shown on a screen at any one given time.

Examples of these interface types are shown in Figure 9.3. Some of the most widely used GUIs are the Apple Macintosh interface, Windows 3 for the PC and MOTIF on X-Windows. Some graphics standards such as the graphic kernel system (GKS) and the computer graphics metafile (CGM, ISO 8682 1987) have been defined to help standardise interface development. Arnstein (1992) provides a comprehensive review of GUIs for GIS, in particular toolkits for the development of GUIs in X-Windows.

9.4 User interfaces for decision support systems

The use of spatial decision support systems has already been discussed in the context of achieving an integrated geo-information strategy for the purpose of problem solving (Chapter 1). A key issue in achieving this is the development of suitable interfaces for the decision maker.

Armstrong *et al.* (1991) recognised the need for the user interface to be flexible enough to present multiple views to the user. For example some information is readily viewed as a distribution on a map, while other data are better presented in a tabular or graph form. A requirement of a SDSS is the ability for the decision maker to be able to access a variety of information, to view this information in juxtaposition if required, to make changes to elements of the information and to view the effects of those changes in real time. This implies a high level of screen organisation and the processing power to perform such tasks.

The design principles that apply to interfaces for SDSS were discussed by Tognazzini (1990), Turk (1990), Raper *et al.* (1990) and Armstrong *et al.* (1991). They include:

- The need for appropriate real world metaphors to be used in organising the graphical environment. The notion of a 'desktop' with wastebasket and folders is familiar to any user of Apple Macintosh computers. It is important that users unfamiliar with the interface can use intuition to infer the purpose of particular graphic objects.
- The interface must be flexible enough to allow users to recover from unintended actions, even after a number of subsequent processes have been carried out. In the Apple environment all files tagged for deletion are kept in the wastebasket until such times as the user chooses to empty this by confirming their deletion. In this way files may be recovered long after their apparent disposal.
- The screen layout should be ergonomically designed so that frequently used tools are easily available and the user has several strategies for dealing with the problem of data or tools hidden by overlapping windows or icons.
- Users should be aware of the processes they are carrying out and the interface should indicate exactly what is the current status of files, applications or tools. The use of colour changes

1. Command line

C: cd \F:/GIS/DEMO/MAP.TXT

2. Fixed screen menu

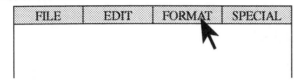

3. Pull-down menu

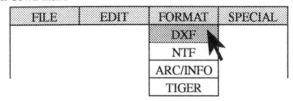

4. Pop-up menu

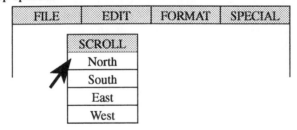

5. Icon

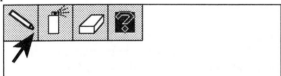

Figure 9.3 Example of generic interface types.

to shown active tools or animation in icons to indicate that some processing is occurring are examples. It is important that these indicators are consistent across applications if the user is not to be confused.

9.5 GIS-independent user interfaces

Rhind *et al.* (1989), Raper *et al.* (1990) and Raper and Bundock (1991) discussed the concepts and development of a GIS-independent user interface environment. The Universal Geographic Information eXecutive (UGIX) contained three main modules:

- Screen interfaces, dialogues and a command processor which utilised Hypertext techniques (see Chapter 12).
- A GIS-independent help system created in a hypertext environment. This was based on the GIS tutor developed by Raper and Green (1989).
- An expert system shell or high level system access module.

UGIX has been implemented for ARC/INFO (Raper, 1991) as HyperArc. It uses Hypercard for the Apple Macintosh where 'the Hypercard application (complete with in-built communications software) acts as a client to a host processor running the GIS application.' (Raper *et al.*, 1990). The use of standardised interface objects makes the use of GIS less daunting for the less technically aware user. The UGIX model allows the various options available at any given point to be displayed and provides pop-up explanations for commands. The interface has four different areas of activity:

- Introduction and explanation (using a map guide)
- Session screens for command processing
- Help environment
- Library for maps and images generated in the GIS.

Most users of GIS require a thorough understanding of spatial theory to make full use of the available functionality. An important aspect of the UGIX model is the help system, which is not a system guide but a guide to the theory and concepts of spatial data handling. It invokes text, graphic and animation techniques in an interactive way to overcome traditional reluctance to utilise system manuals and help facilities. This is based on the Hypercard GIS Tutor (GIST), see section 9.8.

However, it is not necessary to always use Hypercard for this type of GIS-independent interface. The importance of the UGIX model lies in the notion that it is possible to develop a generic interface for spatial information systems. The use of common interface models, such as consistent commands or icons, across systems will require considerable standardisation to be effective.

9.6 Specific applications of graphical user interfaces

There have been examples of user interfaces developed for specific problems. McMaster and Mark (1991), for example, describe the design of a GUI for knowledge acquisition in cartographic generalisation. Their user interface was designed to monitor the activities of trained cartographers as they generalize a map. The GUI was considered the first step in developing automated generalization software. The key features of this interface were:

- Providing the cartographers with a full array of generalization tools;
- A focus on the sequential and interactive effects of generalization;
- Keeping a detailed history of the activities involved in creating a generalized image.

An important aspect of this type of interface is the need for visual feedback to the user and the direct manipulation of the image so the cartographer can assess the results of a particular process (Weibel, 1991). Another example is the development of an interactive interface and a user-programming language for a prototype application for testing thematic map symbolization (Taketa, 1992). The development of the prototype encompassed the basic elements of interface design:

- Easy to use;
- Modifiable, such as the moving of items across menus and within menus;
- Extendable, with easy to add interface elements.

A user interface for decision support in natural resource management was described by Elmes (1991). GypsES is a knowledge-based spatial decision support system for the management of gypsy moths in forest environments. The user interface 'integrates several features considered to be valuable for the successful adoption of the technology in managerial environments having little or no previous GIS or computing experience'.

The system uses a hierarchical series of flow charts to guide the inexperienced user through the most efficient use of the system, while at the same time experienced users have tools to maximise system accessibility. The interface includes:

- The use of colour to indicate current location and 'depth' within the system;
- Navigation aids and reminders;
- Attention-demanding messages which rely current status or user errors;
- Consistency in the interface structure between system components.

These very different examples might suggest that the goal of a GIS-independent interface would be difficult, if not impossible, to achieve. The problem for system suppliers is how to build a general purpose interface that at the same time meets the needs of specific user groups. This implies a requirement for a general purpose interface model for spatial data which can also be customised to some degree.

Currently, none of the widely used GUIs specifically support GIS functionality; for example, there is no co-ordinate display option. Arnstein (1992) notes, however, that most GUI environments are extendable and GIS tailored programming toolkits could be developed.

9.7 Issues of communication

Interface customisation tools would allow integrated solutions to have various access and operational levels within an interface. The customisation would allow individual users to select their own operating strategy, choosing between menus, icons or commands, allow certain features to be placed in screen positions where they act as the necessary cognitive cues, and to change levels as the user passes from familiar to unfamiliar applications.

The design of suitable interfaces for spatial query, manipulation and analysis has to draw together issues of system-user communication in order to optimise use of the system by the user community. There are a number of issues that have to be addressed:

- *Spatial language*. The development of a spatial interface language is fundamental to the development of spatial information systems. This was recognised by the adoption of spatial languages as a key research initiative by the US National Center for Geographic Information and Analysis (NCGIA). Use of the current SQL standard is generally regarded as inadequate, with the need for a geographic query language that uses generic spatial operators such as distance. This language has to address the various levels of user understanding that exist from novice to skilled specialist.

- *Intuitive interface metaphors.* There is the need to develop interface metaphors for spatial information, similar to the 'desktop' analogy used in the Apple Macintosh. These may be developed at a system independent level or at an application specific level.
- *Interface visualisation.* The actual design issues relating to screen layout, colour, object types, interactive changes to objects and use of intuitive icons has to be carefully considered. There is a tendency to develop cluttered interfaces in which it can be difficult for the inexperienced user to locate specific icons. This arises when the interface designer wishes to provide much of the system functionality on a single screen.

The development of interfaces that contain standardised elements that are system independent and customisation tools will be required to effect greater integration for spatial information systems. Much more research is required into aspects of human–computer interaction for spatial data handling.

9.8 Help facilities

Raper *et al.* (1990) highlighted the need for appropriate help facilities as an integral part of the interface. The help facilities can be divided into two types:

- System specific help related to command structures and processing sequences
- General spatial information concepts

The level of system specific help facilities varies from simple command manuals to interactive command prompts and explanation. Many packages come with system tutorials, demonstrations and exercises. The quality of these varies, but for the inexperienced GIS user, they are often inadequate. Suppliers usually have to provide support facilities to help users develop their applications. This is particularly necessary where the basic knowledge of GIS concepts is limited.

For most GIS users there is a need to develop some basic

understanding of concepts and how to apply them before they are able to fully utilise a system. Many training and awareness courses are run to provide the user community with this base knowlege, and this will grow with the introduction of GIS into higher education establishments and schools (Green, 1992; Palladino, 1992; Vicars, 1992; Wood and Cassettari, 1992).

Even with this increase in spatial awareness there will remain the need for help facilities that provide the non-specialist user of spatial information with background concepts. The GIS Tutor (GIST) is an example of this (Raper and Green, 1989). Developed using Hypercard for the Apple Macintosh it is structured much like a book, with 'chapters', index and contents page. It provides a graphical 'map' of the tutor's structure, allowing the user to browse. It also contains spurs, loops and links which allow additional information to be added so the user can develop a particular subject area to a greater degree of detail or cross refer to other sections. The system contains text, graphics and animations.

The integration of this type of GIS tutor needs to be linked with system help facilities to provide a comprehensive help environment. This is increasingly important in the development of integrated, multi-user systems with a broad range of user capabilities.

References

Albaredes, G. (1992) GIS usability in question. *Proceedings 15th European Urban Data Management Symposium,* Lyon, pp. 435-41.

Armstrong, M. P., Densham, P. J. and Lolonis, P. (1991) Cartographic visualization and user interfaces in Spatial Decision Support Systems. *Proceedings GIS/LIS,* Atlanta, pp. 321-30.

Arnstein, D. (1992) Windows in the open. In: *Geographic Information. The Yearbook of the Association for Geographic Information.* Eds. Cadoux-Hudson, J. and Heywood, D. I., Taylor and Francis, London, pp. 449-58.

Egenhofer, M. and Frank, A. U. (1988) Designing object-oriented query languages for GIS: Human interface aspects. *Proceedings Symposium on Spatial Data Handling,* Sydney, pp. 79-96.

Elmes, G. A. (1991) Design and implementation of a user interface for decision support in natural resource management. *Proceedings GIS/LIS,* Atlanta, pp. 399-406.

Goldberg, A. and Robson, D. (1983) *Smalltalk-80: The Language and its Implementation.* Addison-Wesley, New York

Green. D. R. (1992) GIS education and training: Developing educational progression and continuity for the future. In: *Geographic*

Information. The Yearbook of the Association for Geographic Information. Eds. Cadoux-Hudson, J. and Heywood, I., Taylor and Francis, London, pp. 283-95.

HMSO (1987) *Handling Geographic Information.* Report of the Government Committee of Enquiry chaired by Lord Chorley. HMSO, London

McMaster, R. B. and Mark, D. M. (1991) The design of a graphical user interface for knowledge acquisition in cartographic generalization. *Proceedings GIS/LIS,* Atlanta, pp. 311-320.

Mark, D. M. and Frank, A. U. (1989) Concepts of space and spatial language. *Proceedings Autocarto 9,* Baltimore, pp. 538-56.

Palladino, S. (1992) GIS and secondary education in the United States. In: *Geographic Information. The Yearbook of the Association for Geographic Information.* Eds. Cadoux-Hudson, J. and Heywood, I., Taylor and Francis, London, pp. 304-9.

Raper, J. F. 1991 Spatial data exploration using hypertext techniques. *Proceedings European Geographical Information Systems Conference,* Brussels, pp. 920-8.

Raper, J. F. and Bundock, M. (1991) UGIX: a GIS-independent user interface environment. *Proceedings Autocarto 10,* Baltimore.

Raper, J. F. and Green, N. P. A. (1989) Development of a hypertext tutor for geographical information systems. *British Journal of Education Technology,* Vol 3, pp. 164-72.

Raper, J. F., Lindsey, T. K. and Connolly, T. (1990) UGIX - a spatial language interface for GIS: concept and reality. *Proceedings European Geographical Information Systems Conference,* Amsterdam, pp. 876-82.

Rhind, D. W., Raper, J. F. and Green, N. P. A. (1989) First UNIX then UGIX. *Proceedings Autocarto 9,* Baltimore, USA pp. 735-44.

Taketa, R. (1992) Object-based user interface structures: Experiences from a software prototyping application. *Proceedings GIS/LIS,* San Jose, pp. 757-66.

Tognazzini, B. (1990) Consistency. In: *The Art of Human–Computer Interface Design.* Ed. Laurel, B., Addison-Wesley, Reading, pp. 75-8.

Turk, A. G. (1990) Towards an understanding of human–computer interaction aspects of geographic information systems. *Cartography,* Vol 19, No 1, pp. 31-60.

Vicars, D. (1992) GIS and higher education. In: *Geographic Information. The yearbook of the Association for Geographic Information,* Eds: Cadoux-Hudson, J. and Heywood, I., Taylor and Francis, London, pp. 310-312.

Weibel, R. (1991) Amplified intelligence and rule-based systems. In: *Map Generalization: Making Decisions for Knowledge*

Representation. Eds. Buttenfield, B. P. and McMaster, R. B., Longman, London.

Wood, A. and Cassettari, S. (1992) GIS and secondary education. In: *Geographic Information. The Yearbook of the Association for Geographic Information*, Eds. Cadoux-Hudson, J. and Heywood, I., Taylor and Francis, London, pp. 296-303.

10

Visualisation

10.1 Cartographic visualisation

The origins of spatial information systems are partly to be found in digital cartography. The desire to enhance or even replace the manual work of the cartographer with digital solutions that replicated the map making process consumed much research and development time in the 1970s and 1980s (Boyle, 1982; Jackson, 1982). For the large map production agencies the process of conversion from a manual to a totally digital solution is still of prime importance and will remain so until large scale digital map databases exist. The Ordnance Survey will have converted the 1:1250 and 1:2500 paper maps which cover much of the UK into digital form by 1995, but on a global perspective this is an exception rather than the rule.

These origins have been significant in determining the type of cartographic presentation found in GIS. Most systems aspire to provide the user with the tools to reproduce the paper map product as accurately and as faithfully as possible, since many users are happiest with digital maps that look like the paper maps with which they are familiar. This raises many issues about how we use maps in a GIS and their effectiveness as a communication tool. Without the appropriate communication methodologies there is little point in establishing spatial information systems for the purposes of broad decision making.

The importance of good cartographic design has been widely discussed for analogue maps and there have been some recent attempts to formalise digital map specifications and to define the differences between paper maps and screen images. The generally held view of the GIS user is that cartographically acceptable paper products are equally acceptable in digital form. This is not necessarily the view of cartographers and a number of studies, such as Gooding and Forrest (1990), have demonstrated that digital maps are not as readily interpretable as the paper form of the same map.

In this chapter we discuss the issues relating to cartographic design and the wider field of visualisation in the context of GIS. The study of visualisation encompasses the generation, manipulation and comprehension of the products of vision (Buttenfield and Mackaness, 1991). From the cartographer's perspective, visualisation involves the visual processes that allow us to identify patterns and to create and manipulate mental images. This cognitive aspect underlies much of the research in map interpretation but visualisation extends to the human–computer interface and includes the interactions between computation, cognition and graphic design (Buttenfield and Mackaness, 1991).

10.2 Perception and patterns

The history of cartographic development has been about the methods of presenting information on spatial location and at the same time conveying a broad 'understanding' or 'impression' of the environment being depicted. Increasing sophistication in surveying techniques and in the methods of graphic representation has enhanced the spatial accuracy of the data and the quality of their depiction. This may be seen in the move from latitude and longitude as measures of spatial location to alternative grid co-ordinate systems or in the replacement of drawings to represent relief features by a more formal representation such as contours. Figure 10.1 is an extract from one of the early maps of London. The purposes of such surveys were to provide a basic inventory of what existed and where, and as part of this to identify what existed in relation to what. In the production of such maps a large amount of cartographic licence was used in converting the original surveyors' plans to the final product and yet the result conveys the general tenure of the landscape even if it lacks accuracy by current standards.

This increased formalisation in the way maps depict geographical objects has increased the need for greater cartographic awareness by the map reader in order to extract the appropriate information. Many people find relief information is more difficult to interpret from contours than from certain methods of hillshading. The result is that training and education into the use of maps and what they can tell the geography student have become part of the school curriculum in most countries. The growth in GIS has only served to re-emphasise the need for knowledge and awareness about the role of maps as a geographical inventory.

There also remains an important role for maps in providing a 'perception' or an 'image' of the environment. An example of this is the use of maps for making political statements about sovereignty or boundary disputes. In such cases spatial accuracy may be one of the least important considerations. Another example is the use of pseudo

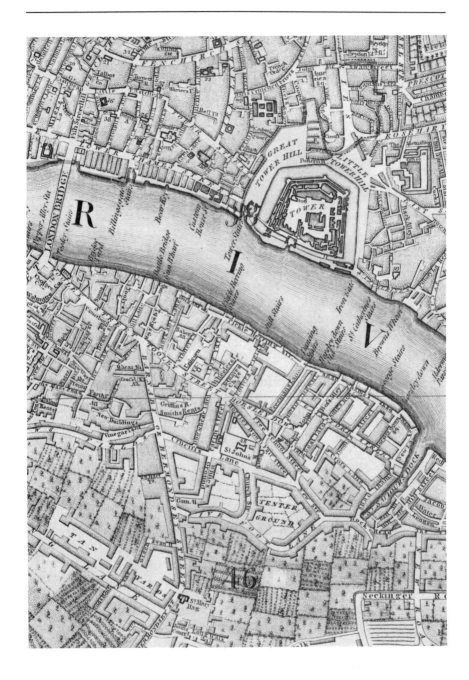

Figure 10.1 Extract from Cary's map of London published in 1799.

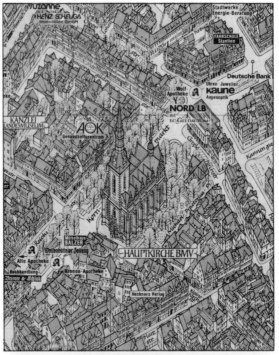

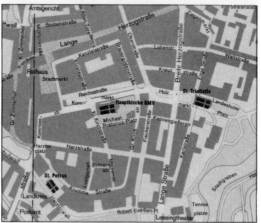

Figure 10.2 Differing approaches to the representation of town plan information.

perspective views of buildings on a town plan to convey the general character of the town. The example in Figure 10.2 is for the town of Wolfenbuttel in Germany, which compares this technique to the more traditional approach used in street plans.

The other key use of maps is for the recognition of patterns and for defining homogenous areas based on particular criteria. The importance of these has increased as the map has moved from being purely an inventory tool for recording the existence of features to a more analytical tool used for identifying such patterns. This has resulted in the development of a branch of cartography which deals with thematic maps designed for the purposes of highlighting patterns in the features themselves or the processes which form their interaction. An early example of this type of mapping was the study undertaken in London in the 1890s to identify the extent of poverty.

Thematic mapping has become an important output from GIS and is often the most tangible product of an analysis. There is an increasing use of simple thematic maps to portray points of interest in newspapers and on the television, while many of the low end computing GIS packages simply provide the graphic tools and associated database for generating thematic maps. Packages like GIMMS are good examples of systems designed for the purposes of representing data from sources such as a national census and combining the map output with graphs and diagrams. Figure 10.3 is an example of a map generated from the UK's 1981 census.

The problem for most spatial information systems is that little thought has been given to the map design component of the GIS. Computer graphics specialists have been involved in providing a wide array of drawing tools, colours, charts and the means to create new symbols. However, few cartographers have been involved in establishing appropriate digital methodologies for the map creation process and hence it is very easy to generate maps that convey a false perception of the features they depict or do not allow the user to identify the patterns which exist within the data. The example in Figure 10.4 shows how the use of inappropriate shading symbology on a thematic map obscures the existence of a pattern or trend.

10.3 Map design issues

Cartography has been described as both a science and an art. There is an element in the creation of a map which involves the cartographer in exercising both personal judgements and preference in the way information is portrayed. A large measure of this personal element is removed for series mapping by the creation of standardised map

PENSIONERS

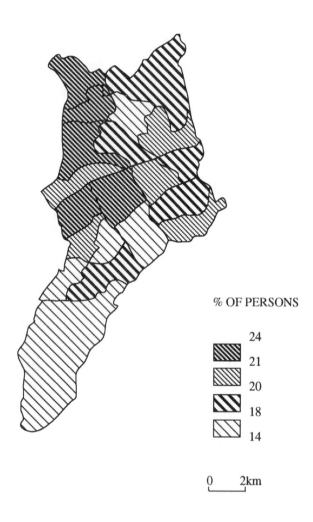

% OF PERSONS

24
21
20
18
14

0   2km

Figure 10.3 Distribution of pensioners by ward for Kingston-upon-Thames based on 1981 census data.

specifications which dictate when and where symbols should be used, which features should be placed in their correct position and which should be displaced and what colours, type styles and line weights are used. Even so, a look at the different 1:50,000 topographic map series produced by the national agencies across Europe reveals a wide range of styles (Cassettari, 1989). This is in part a function of the nature of the terrain being portrayed, for example the representation of relief on the Dutch series is much less dominant than on the Swiss series, but it is also a function of map design.

The variations in these styles of map raises the question of what constitutes a good map design. The academic study of map design did not emerge as a scientific discipline until after the Second World War when mapping for military purposes had achieved a global level of importance (Buttenfield and Mackaness, 1991). Even so Robinson (1975) noted that there has been 'little or no ordered examination' into the design of maps even though it is a fundamental element in the communication of cartographically presented information.

Wood (1968 and 1972) noted that the design of a map has to take into account the problems of visual search and recognition, the role of memory in interpretation and the user's own reality in comparison to the cartographer's reality. Figure 10.5 shows the relationship of the cartographer to the map user in the context of the depicted and perceived realities of the environment the map depicts.

The basic interactions between the map designer and the map user are just as important in a digital environment, where a knowledge of traditional cartographic design principles should be applied to the design of digital maps. However, while the development of digital mapping systems has resulted in some innovative graphical techniques which would be tedious or impossible to achieve without computer assistance, there has been little concern in cartography for designing new methods to extend what is possible on the map (Buttenfield and Mackaness, 1991).

Moellering (1980) recognised that digital maps are in many respects different from their paper predecessor when he defined the concept of 'real' and 'virtual' maps. This recognised that the 'real' paper map has both a permanent tangible reality and is directly viewable as a cartographic image. Virtual maps on the other hand will have only one or neither of these attributes and can include images which may or may not require further processing to be readily viewable (Moellering, 1983).

This recognises that the digital map is a cartographic interpretation of the digital database drawn onto a computer screen only at the behest of the user and that the map may only be on the screen for a matter of seconds before it is lost to the viewer, perhaps forever.

The practical issues of what constitutes a good map can be

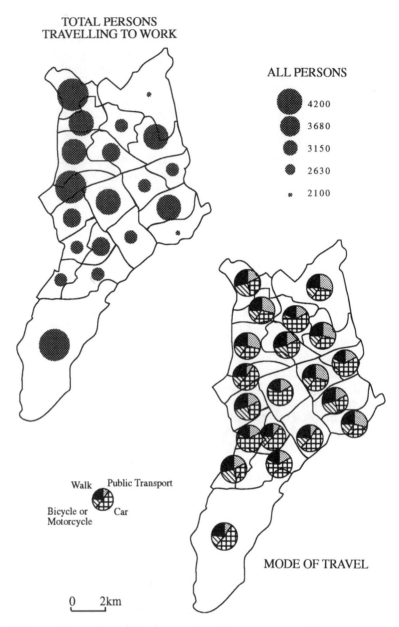

TOTAL PERSONS
TRAVELLING TO WORK

ALL PERSONS

4200

3680

3150

2630

2100

Walk   Public Transport

Bicycle or   Car
Motorcycle

MODE OF TRAVEL

0   2km

Figure 10.4   Thematic maps for Kingston-upon-Thames using proportional circles and pie charts to represent variations in categories of information from the 1981 census.

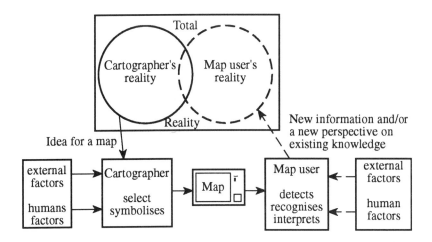

Figure 10.5   Cartographic communication (after Wood, 1972).

summarised from a debate initiated in the British Cartographic Society which defined the following areas for continuing discussion:

- Colour and how it is used both as an indicator for certain feature types, for example water features are always blue, and to highlight specific feature attributes, such as the use of purple in some atlases to denote very high relief features. Understanding the use of colour is very important to the impact of the map but is seldom well used in a digital environment.
- Symbology and the move towards more standardised symbols such as those found on tourist maps. The intuitive nature of symbols and the problems of designing the universally understood map symbol has been of increased interest with the international nature of map publishing, for example studies into the interpretability of tourist symbols on maps.
- Typology and the use of varying text styles and fonts on a map to highlight features, the problems of text placement, alignment and overlap with other map detail. These are of particular concern in a digital environment where there is less control over text placement and in many cases the GIS output contains text information which is partly obscured by other features.

These components of a map are briefly considered in the following sections.

10.4 Colour and symbolisation

The issues of colour and symbolisation have been of growing concern for those interested in maintaining the quality of cartographic design in a digital environment. The trend amongst UK cartographers who are part of the independent cartographic publishing industry is to use the Apple Macintosh platform for cartographic compilation, making use of the varied and sophisticated graphics packages for compiling maps. Such packages offer the cartographer a choice of millions of colours, a wide range of pens, brushes, paints fills and sprays with the ability to create new shade fills and symbols as desired. For those who become adept with the tools the flexibility and variety of techniques provides a greater opportunity to introduce an artistic element into the cartographic process. In these circumstances it is the professionalism of the individual cartographer which ensures the continuing adherence to good cartographic design.

What is of growing concern to the cartographic profession is the creation of maps by those with little or no cartographic training, which results in a map which is either difficult to interpret or even worse conveys the wrong impression. In Figure 10.6 the symbols used in the map reprinted in a newspaper convey the wrong impression to the reader about the areas with the highest levels of crime. It is even easier to use colour to convey the wrong visual impression.

The traditional idea of using fewer colours to depict any combination of areal units on a thematic map, or the idea of using the most striking colours to depict the greatest or most significant category of information are tenets not always adhered to. For the information manager the increasing availability of colour is potentially a very powerful communication tool made all the more powerful by the widespread availability of colour reproduction and copying devices. However, the poor use of colour can lead to problems of image synthesis where the viewer uses image elements to reconstruct the full scene in a way which leads to problems of interpretation.

10.5 Typology and text placement

Much effort has been put into the automation of the typographical components of maps for the purposes of generating digital map productions which conform to the same specifications as paper maps. To this end sophisticated graphics packages provide tools for the

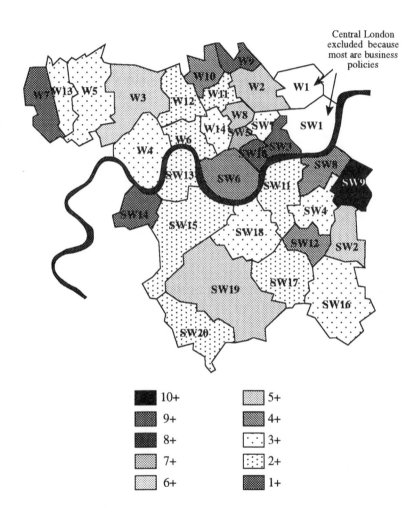

This map shows the risk of a break-in at
your home in percentage terms.
The national average is 3.1%

Central London
excluded because
most are business
policies

■ 10+	⬚ 5+
■ 9+	▨ 4+
■ 8+	⬚ 3+
□ 7+	⬚ 2+
□ 6+	▨ 1+

Figure 10.6 Extract from a map published in a newspaper which
conveys the wrong impression to the reader

handling of text which allows interactive font and size changes, format changes, text rotation and the ability to 'lay' text along a curved or sinuous feature such as a river. While these are important tools where the typology is well defined, it is equally important to address the issues of the most appropriate text font, size and style for a particular purpose if those involved in compiling maps are to make appropriate use of this facet of cartographic compilation. An inappropriate use of text can destroy the visual impact of graphic image.

Most GIS solutions do not offer sophisticated graphics capabilities and even poorer text management systems. The example of ARC/INFO is a case in point. The standard text font is quit distinctive and rather simplistic. While it is possible to use alternative fonts or even to create your own their use is cumbersome. As a consequence the best results are obtained when map frames containing appropriate text, legends and notes are set up in advance and used as the base on which to display the results of a spatial analysis or query. This is not untypical of many GIS solutions in which many of the map products contain only limited typographical content. Thus much of the text information placed on a map as a means of providing additional spatial references or as keys to interpreting the information is lost. Examples of text used in these ways might be regional names such as 'The Appalachians' or the descriptor to a symbol such as 'Council Offices' or 'Town Hall' next to a significant building.

The alternative to omitting text is to add too much. The use of text panels or free text windows which overlay the map information can be difficult to read and can obscure the underlying map information. Much text placed on digital maps is partially obscured because it crosses inappropriately shaded or coloured areas, making it difficult to read. Some text is poorly placed and the use of hierarchical typology specifications to denote types of features, size or importance are often ignored. Figure 10.7 shows examples of the text fonts and sizes it is possible to achieve very simply in many word processing, desktop publishing and digital cartography applications.

The point to be made is that cartographic representation of an area involves not only graphic symbology for the types of features of interest but also some additional information in the form of text which the map reader assimilates into the overall interpretation of the cartographic product. It is therefore important to include this type of detail in the digital map or if it is to be omitted to ensure the graphic components replace the information provided by the text on a paper map.

There may be some very sound reasons why text is not so easily adapted to a GIS environment:

• Screen size and the problems of information clutter;

AaBbCcDd	**Chicago**
AaBbCcDd	Courier
AaBbCcDd	**Geneva**
AaBbCcDd	Helvetic
AaBbCcDd	Monaco
AaBbCcDd	**New York**
AaBbCcDd	Palatino
AaBbCcDd	Times

1. Various standard text fonts

abcdefghijklmnopqrstuvwxyz   Times 9pt

abcdefghijklmnopqrstuvwxyz   Times 10pt

abcdefghijklmnopqrstuvwxyz   Times 12pt

abcdefghijklmnopqrstuvwxyz   Times 14pt

abcdefghijklmnopqrstuvwxyz   Times 18pt

abcdefghijklmnop..   Times 24pt

abcdefgh.. Times 36pt

2. Various standard text sizes

Figure 10.7 A selection of the text sizes and fonts available for customising cartographic output

- Screen resolution and the problems of visually interpreting different text fonts and sizes;
- The time it takes to place text information, particularly if the map has only a limited screen life;
- Limited text handling capabilities of most GIS packages;
- Poor understanding of the role of text in cartographic products by the non-cartographic map compiler;
- The availability of alternative information communication strategies in an interactive cartographic environment.

It is this last which offers a range of alternatives for those who are not inclined to invest resources into adding text to GIS map output. It is quite possible to use pop-up free text windows, graphic icons, a wider range of colours and symbols, or sound, all of which need only be accessed when required. What this demands is a revised approach to the use of traditional cartographic forms in an information management environment.

10.6 Interactive map specifications

The alternatives to the use of text and the greater range of colours and symbology available for the compilation of digital maps suggests the need for a review of cartographic theory in a digital environment. Robinson (1975), Moellering (1983) and Buttenfield and Mackaness (1991) have implied that existing cartographic theory as developed from paper-based maps is not adequate for an interactive computer environment and that much more research needs to be undertaken into this field.

The use of geographic layers or entity-relationship models for grouping geographical entities and the ability to link information such as text, graphics or sound to the geographical elements in a relatively unstructured form, such as the Hypertext model discussed in Chapter 12, should improve the overall quality of information organisation. These allow the map compiler to adopt a different strategy to the presentation of information which will not depend on all the data being available on the screen in one viewing. Instead the user will be able to access more detailed levels of information as required. This sort of strategy allows the user to compile a map by selecting and ordering information as desired. ARCVIEW is an example of how a user may query a number of ARC/INFO maps or coverages and create a structured map based on an ordered selection of map layers.

The danger with this approach is that there is little control over the type of map the user compiles and thus without appropriate

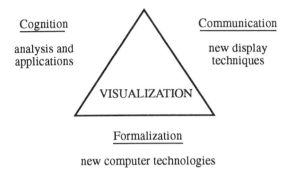

Figure 10.8 Components of visualisation (after Taylor, 1991).

understanding of the cartographic visualisation issues the information that is communicated may be poor or inappropriate. There is clearly a need for greater research into the issues of cartographic design in an interactive environment. The purpose of such research must be to identify the key cartographic, visualisation and communication issues in digital interactive map display in order to define default system standards that provide for high quality map output in the GIS environment.

10.7 Scientific visualisation

The development of new technologies to cope with sophisticated image management has led to the emergence of scientific visualisation. Powerful computer systems are able to provide image manipulation tools that encompass real-time image modelling and virtual displays. The types of systems extend from health care activities, where scanned images of the body are modelled and displayed as three-dimensional images in real time, to architectural modelling of proposed developments. The modelling allows the image to be viewed from all external aspects and to create cross-sections or internal views.

Scientific visualisation is primarily concerned with the relationships between the technology and the aspects of cognition and communication that enable it to be used successfully in a problem solving context. It involves various techniques of analysis and display in a technically precise way. Figure 10.8 shows the relationship of these aspects, identifying the new computer technologies as the formalism component of the visualisation model.

There is a need to understand the interaction of the three visualisation components to the successful utilisation of spatial information management tools (Cassettari, 1993). As the GIS or SDSS becomes more integrated with other systems within the decision-making structure, so the importance of visualisation increases. The competing issues of accuracy, quality and context can all be viewed in different ways depending on the nature of the spatial information system but over-riding these are the cognitive and communication elements which are often ignored in system design and data presentation. The inability of cartographers to establish design criteria and visualisation strategies for digital maps is likely to lead to poorer maps in GIS and SDSS and a consequential loss in information quality.

10.8 Terrain representation

The previous discussions have focused on the cartographic visualisation of two-dimensional geographical data but the visualisation of the third spatial dimension provides another level of visual enhancement for the decision maker. Many GIS packages contain functionality for the representation of relief data in the form of perspective views and surface models based on profiling. Figure 10.9 is an example of a simple perspective terrain view where the elements created by profiling in either the x or y direction create an impression of the land surface. This can be enhanced by exaggerating the vertical dimension. In examples where profiles in both the x and y directions are combined the resulting cells can be shaded to indicate such attributes as height, slope or land use. The use of colour for such representations is a very important enhancement. Such representations can only be achieved in a digital environment.

These types of terrain visualisation are 2.5-dimensional representation of surfaces and are distinctive from true 3-dimensional representations, discussed in the next section (Raper and Kelk, 1991). A single height value is attached as an attribute to each x,y co-ordinate position thus providing the information for the z-axis.

A number of visualisation techniques for digital terrain data were described by McLaren and Kennie (1991) in which they identified both static visualisation, which is used to communicate results and concepts, and interactive visualisation, which allows the user to explore the terrain models from different elevations, azimuths and distances. In the most developed form the interactive visualisation lets the user 'fly' around the terrain. The techniques of terrain visualisation are summarised as:

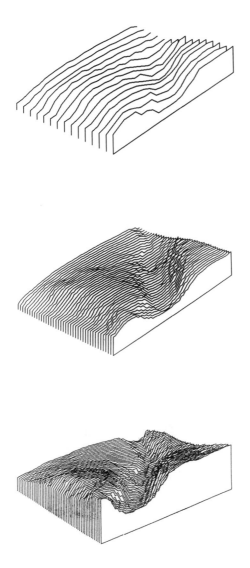

Figure 10.9   Example of a simple terrain surface created using relief profiles.

- *Contouring, including hypsometric tints.* The contour is still the most widely used form of showing relief. As a visual abstraction of the terrain contours remain an effective method of displaying information about the third dimension. Their limitation lies in the users' ability to interpret the contour data and to generate a mental image of the terrain. This is dependent on experience of different terrain types, three dimensional spatial awareness, visual acuity and understanding of how contours are used to depict terrain. Alternatively the use of coloured relief layers or hypsometric tints to denote changes in height has been widely used for paper maps and may equally be applied to GIS maps. The three dimensional visualisation may also be improved by other techniques such as inclined contours and shaded contours (Poiker, 1982 and Yeoli, 1983).
- *Hillshading.* The depiction of relief by shading the terrain according to the slope angle and position of a fixed light source has been widely used on paper maps and has become a standard display technique in many GIS solutions. Its main limitation is the lack of actual relief values so it is possible to generate a visually exaggerated impression of the relief.

    The process of calculating a hill shading model involves the calculation of the surface normals, or surface slope, for each point on the DTM and then the application of an illumination model. It is usual to be able to alter the model to depict the terrain at different times of the day and with the light source in different locations. The usual convention is that adopted for paper maps with the light coming from the north-west. A number of illumination models exist, ranging from the simplest Lambertian shading to more sophisticated techniques such as Gouraud shading and Phong shading.
- *Perspective views.* The disadvantage of the previous types of terrain representation are that they do not provide a visually accurate impression of the surface. A better approach is the use of perspective views which provide a better relative spatial context of relief features.

    The problems with generating perspective views are the three dimensional transformations necessary and the method of calculating the 'hidden' elements of the surface. The simplest form of perspective views uses a wire net technique, similar to that in Figure 10.9, while more sophisticated systems incorporate automatic hillshading into an interactive view generation process. A particular type of perspective view is the panoramic view which places the observer on the ground and is used in studies of intervisibility.

The problem with this type of terrain visualisation is the computer processing required for the more sophisticated models. In the most sophisticated solutions a hardware element is used to build the most sophisticated displays.

- *Surface overlay.* A further enhancement of the terrain surface is to combine the DTM with other datasets to improve the visual quality of the depicted surface. A common technique is the drapping of raster data over a DTM, in particular orthophotos and remote sensing images have been used for very effective visual displays of terrain (Figure 10.10). A further enhancement is the replacement of shaded pixels by standard symbols or pictograms such as buildings or trees. These may be scaled to fit within the perspective view.
- *Animation.* A final aspect to this type of visualisation is the use of animated sequences of terrain views to give the impression of moving around the environment. The large volumes of data and the processing speeds means that animations are the most effective form of 'pseudo travel' across a terrain environment. A good example is that generated for part of Mars from data collected from the Voyager missions in the late 1970s.

10.9 Three dimensional visualisation

The 2.5-D visualisations are important for many application areas because they afford a view of the terrain similar to that experienced by the user. There are however a number of application areas where true three dimensional visualisation is a necessary part of the manipulation and analysis process. Raper (1989) discussed the development of a conceptual design for three dimensional spatial objects. An important distinction between these types of data and the two dimensional or 2.5 dimensional data are the lack of detailed knowledge about the three dimensional world, particularly in the subsurface studies. For example, a three dimensional model of geological strata may be created from data interpreted from borehole cores.

The data model and data structure are clearly important in being able to manipulate and analyse such data, as discussed in Chapter 3. But equally important are the ways in which the data are presented. Three dimensional visualisation involves the display of a true three dimensional object from any user-selected location. Typically this will be external to the object from a specified direction or distance, but can also include the display of analytical operations such as a cross-section through an object or the results of intersecting spatial objects.

True three dim interactive visualisation is not easy to

Figure 10.10 Example of a digital perspective terrain view with a SPOT remote sensing image overlain on a digital elevation model.

achieve. Movement of the observer's position around an object or zooming in and out to show increasing amounts of detail require considerable computer resources. The types of spatial models developed for use in aircraft simulators have to be able to present suitably accurate three dimensional images of an airport and its surrounds which are responsive to the changes in aircraft position resulting from the simulated flight path. Such models also integrate animation with the spatial data to add an extra level of realism, such as the movement of vehicles around airport buildings as they would appear on the periphery of a pilot's vision.

The traditional approach to three dimensional data is to utilise cartographic modelling techniques to visualise the information. The development of techniques that allow the user to replace the cartographic representation with a true image of the feature or a stylised image represents a move away from cartographic representations to image-based systems. These are discussed further in Chapter 11.

10.10 Perspective cartography

One approach adopted by traditional cartographers to the problem of three dimensional visualisation was to generate perspective views of particular features, especially mountainous areas which had interest as tourist locations. These images embodied cartographic components with the representation of roads and other features in symbolised and stylised forms as well as a more artistic representation of the terrain. Such perspective 'maps' are still popular since they require less interpretative skills than traditional maps to extract information about the essential geographical context of features.

The development of sophisticated graphic tools for use in the design of perspective images which have a known geographical location provide an opportunity for the cartographer to generate 'perspective maps' in a digital cartography or GIS environment. For example it is a simple process to design a symbol that represents a building as it is seen from ground level. This symbol may then be used to represent a building or farm and placed on a terrain model adding a further element of interpretability to the image. A single symbol can been repeated and overlaid to represent a larger built-up area such as a village, and the size of the symbols can been adjusted to fit with the perspective view of the terrain. Further realism is achieved by ensuring that symbols which would be obscured from the user's vantage point are not shown or are partially visible. The use of a few simple symbols representing a building, tree, road and river are very powerful in providing an enhanced visualisation of the area.

Stage	Input	Analysis	Output
First	ad hoc, in-house digitisation	none; stores and retrieves digitised maps	hardcopy; goal is to replicate existing products
Second	centralised data capture, data exchange	single-state analysis, static modelling	interactive softcopy graphics; successful replication of existing products
Third	incremental updates, dissemination of d change data	multi-state analysis, ynamic or predictive modelling	animated graphics multi-temporal maps, new product designs

Table 10.1 Three stages of GIS capabilities (after Langran, 1991).

This use of pictorial symbols in a perspective cartographic representation is an important development in the broad communication aspects of visualisation. The development of very sophisticated graphic symbolisations of features occurs in computer games because there is a need to engender a degree of realism for the game player. Cartographers should be able to utilise such developments and create maps that are essentially perspective digital cartographic products.

This visualisation model may be extended by the use of real images rather than symbolised images for specific features such as notable buildings, bridges or monuments. The idea of bringing together images and graphic representations of buildings has been used extensively in architectural planning and design (see Chapter 12).

10.11 Visualisation of temporal information

The consideration of time as a component in GIS has increased significantly with the awareness that temporal change is an important aspect of spatial modelling. Langran (1991) noted that the 'ability to work with time could herald a new stage of GIS'. Table 10.1 shows the three stages of GIS capabilities identified by Langran with the output becoming increasingly complex. One of the major functions of a temporal GIS is that of display with the ability to generate a static or dynamic maps, or a tabular summary, of temporal processes at work in a region.

In most digital mapping or GIS systems the only temporal components that exist are information about the currency of spatial datasets such as compilation dates, revision history, when digitised and so on. While it is extremely important for the user to know this in order to appreciate some of the limitations with the data this type of data is rarely displayed. Most screen maps do not give details about the map in the way that they appear on printed maps, albeit in small print and often in a coded form.

Time series data may be recorded as attributes of spatial entities. Statistical summaries showing temporal change such as those derived from census returns or unemployment records may be associated with areas. These types of data are simply displayed and interpreted using existing cartographic forms such as the thematic or chloropleth map.

More importantly, however, is the ability to display information on changes that occur in spatial objects through time. A common problem is how to show changes in administrative boundaries and the impact of such changes on statistical calculations for areal units which are supposedly the same. There is also the problem of showing gradual change, such as occurs in the position of a river channel. Temporal maps of these changes can be developed into geo-referenced animations. The visualisation of such 'temporal maps' is not well understood and requires considerably more research to identify successful methodologies for successfully communicating temporal information.

10.12 Virtual reality

A final aspect of visualisation which is currently being developed is that of virtual reality. This is where the use of computer images projected onto small screens within a headset, together with technology for simulating touch and sound are integrated to present the user as a virtual world in which their own responses are part of the world. Hand movements in front of the face are seen on the computer screen, turning the head changes the display and apparently touching objects in the computer model leads to touch stimulations which make the user believe the object has been grasped.

This type of technology is becoming increasingly popular for games since it allows the user to become an integral player within the game rather than an external controller. For example the player controls the sword and traverses the corridors of a mythical castle in the search for the solution to a mystery.

This type of technology utilises computer graphics and animations of the virtual world for the user. If however, these were replaced with spatial data, it would be possible to recreate real world locations for the decision maker who may be working anywhere else in the world. The limitations of such solutions will be the quality of the data, the computing power required to handle true spatial data rather than compute generated graphics and the ability of the user to interpret the information. This type of problem highlights the need to address the way people interpret images and compare the visual record with their own experience in order to interpret that record. The use of different visualisation techniques in cinemas can have the effect of persuading the audience that they are somewhere where they are not and the range of emotions associated with the interpretation of the image can be extreme.

This extension to the visualisation concept needs careful consideration because it is easy to make use of technical 'wizardry' for the purposes of achieving the 'golly' factor (Parsons, 1992). This is simply an expression resulting from the immediate impact of the computer images but the user is no better equipped to interpret them and therefore to improve the decision making process. This is an important lesson that GIS users must learn from the experiences of the remote sensing and multimedia communities.

References

Boyle, A. R. (1982) The last ten years of automated cartography: A personal view. In: *Computers in cartography.* Eds. Rhind, D. and Adams, T., British Cartographic Society, London, p1-3.

Buttenfield, B. P. and Mackaness, W. A. (1991) Visualisation. In: *Geographical Information Systems Volume 1: Principles.* Eds. Maguire, D. J., Goodchild, M. F. and Rhind, D. W., Longman, London, pp. 427-443.

Cassettari S. (1989) Scaling the heights. *Geographical Magazine,* Vol. LXI, No. 8, pp. 34-7.

Cassettari S. (1993) Visualisation: Cartography to virtual reality. *Proceedings Survey and Mapping Conference,* Keele, pp. 367-78.

Gooding, K. and Forrest, D. (1990) An examination of the difference between the interpretation of screen based and printed maps. *Cartographic Journal,* Vol. 27, No. 1, pp. 15-19.

Jackson, M. F. (1982) Automated cartography at the Experimental Cartography Unit. In: *Computers in Cartography.* Eds. Rhind, D. and Adams, T., British Cartographic Society, London, pp. 133-42.

Langran, G. (1991) *Time in Geographic Information Systems.* Taylor and Francis, London.

McLaren, R. A. and Kennie, T. J. M. (1991) Visualisation of digital terrain models: techniques and applications. In: *Three Dimensional Applications in Geographic Information Systems*. Ed. Raper, J. F., Taylor and Francis, London, pp. 79-98.

Moellering, H. (1980) Strategies for real-time cartography. *Cartographic Journal,* Vol 17, No 1, pp. 12-5.

Moellering, H. (1983) Designing interactive cartographic systems using the concepts of real and virtual maps. *Proceedings Auto Cart 6,* Ottowa, pp. 53-64.

Parsons, E. (1992) The development of a multimedia hypermap. *Proceedings Association for Geographic Information Conference,* Birmingham, pp. 2.24.1-6.

Poiker, T. K. (1982) Looking at computer cartography. *Geographic Journal,* Vol 6, No 3, pp. 241-9.

Raper, J. F. (1989) The 3-dimensional mapping and modelling system: a conceptual design. In: *Three Dimensional Applications in Geographic Information Systems*. Ed. Raper, J. F., Taylor and Francis, London, pp. 11-20.

Raper, J. F. and Kelk, B. (1991) Three-dimensional GIS. In: *Geographical Information Systems Volume 1: Principles*. Eds. Maguire, D. J., Goodchild, M. F. and Rhind, D. W., Longman, London, pp. 299-317.

Robinson, A. H. (1975) Map design. *Proceedings AutoCarto 2,* APRS, Falls Church, pp. 9-14.

Taylor, D. R. F. (1991) A conceptual basis for cartography; new directions for the information era. *Cartographic Journal,* Vol 28, No. 2, pp. 213-16.

Wood, M. (1968) Visual perception and map design. *Cartographic Journal,* Vol 5, No. 1, pp. 54-64.

Wood, M. (1972) Human factors in cartographic communication. *Cartographic Journal,* Vol 9, No 2, pp. 123-32.

Yeoli, P. (1983) Cartographic drawing with computers. *Computer Applications.* Special issue Vol 8.

11

Image-based spatial information systems

11.1 Image-based information systems

The map is central to the majority of spatial information systems as it forms the basis on which most geo-databases are built. It is, however, possible to carry out many spatial analysis operations which result in text output only. For example a GIS may be used to identify households that will be affected by remedial work to a gas supply main and who will experience an interrupted supply for a period. These affected households may be advised of the proposed work through letters generated automatically based on the address data linked to each household location.

It is conceivable that the use of maps can be misleading to some managers and result in mis-information when it comes to interpreting spatial data within the wider decision-making context. As already discussed in Chapter 10, the quality of the cartographic representation is very important in the ability of individuals to accurately and quickly interpret the information presented to them. This process is dependent on both their visualisation skills and their experience.

This raises the question of what alternatives exist for the map to provide information on spatial context. The developments in storage, retrieval and display technologies have brought together the raster and vector approach to geo-information handling, once viewed as essentially independent. This has led to strategies for integrated handling of maps and image data. In particular remote sensing and digital photogrammetry offer the basis for an alternative to the map-based GIS.

The key feature of this approach is the use of aerial photographs or satellite images as a replacement or enhancement to the map. Traditionally these have been regarded as primary data sources from which maps have been derived. Now they may be used as an alternative to the map, providing base information in image form which may be enhanced by the use of symbology or text. The use of images rather than maps can provide a more readily interpretable view of the environment under consideration. Although the perspective may be unfamiliar for many types of features the image provides details not usually shown on a map.

Furthermore, the extraction of spatial information from such sources need no longer be regarded as requiring specialised skills or computing resources. The increasing availability of imagery in a form suitable for direct data extraction means that a broad range of professionals can use the imagery not only to visualise the environment but also to generate user-specific datasets. This implies a degree of post-capture processing undertaken before the data are made available to the GIS user, such as the removal of scaling errors, the adding of a georeferencing system or the creation of 'true' colour satellite images.

This chapter briefly looks at some of the specific issues relating to these data types and considers the potential for the development of what may be termed image-based spatial information systems.

11.2 Vertical aerial photography

Aerial photographs flown for the purposes of map compilation and revision have existed since the 1920s. Such photography is flown within precise parameters which give carefully calculated overlaps between fore and aft images along a flight line and between flight lines (Figure 11.1). These overlaps between images enable it to be viewed stereoscopically, creating for the user a three dimensional image. This type of aerial photography is known as survey standard vertical aerial photography and until the mid-1980s nearly all such imagery was black and white. Some specialist surveys were undertaken in colour or using thermal or infra red film but these were generally regarded as too expensive for large mapping projects.

At the same time aerial photography has also been used by a wide range of civilian and military agencies for the collection of many types of geographical information, such as land cover surveys, forestry management and land use changes studies.

The development of high resolution colour film and cheaper processing techniques in the 1980s allowed colour aerial photography to be produced at a comparable cost to black and white (Cassettari, 1992a). The introduction of colour film in the UK reversed a downward trend in the commissioning of aerial surveys and, with the growing interest in GIS, led to wider use of aerial photographs.

Black and white photography is still used. The aerial survey of Scotland at 1:24,000 flown in 1987-89 was black and white for the

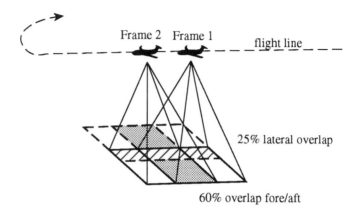

Figure 11.1 The image overlap for stereoscopic aerial photography.

whole country with the exception of the central lowland areas covering Glasgow and Edinburgh, about 20% of the survey (Kirby, 1992).

Aerial photography provides an important source of land use and environmental data not available from the topographic and large scale base mapping. Greater environmental awareness and the increase in legislation on environmental monitoring and protection, together with the need to create more detailed resource inventories, down to the level of individual items of street furniture, has prompted wider use of imagery at scales from 1:25,000 to 1:3,000. Habitat and archaeological surveys, the preparation of grounds maintenance plans and urban modelling are examples of the types of study currently being undertaken in the UK.

11.3 Photogrammetric data collection

Aerial photography has a central purpose as a source of information from which to compile and revise maps. The extraction of accurate spatial information is based on the use of the stereo model inherent in vertical survey standard photography and uses photogrammetric methods for recreating the model in an analogue or analytical environment.

In the UK the availability of large scale mapping means that photogrammetry has a less important role in the collection of data for mapping purposes compared with other countries where the base mapping is not available or out of date. However, even in the UK the Ordnance Survey has a rolling programme of flying to update the base mapping. Photogrammetry is also used for special purpose surveys (Cassettari, 1992b) including the collection of heighting information to complement the 1:10,000 topographic maps and engineering surveys.

In North America the emphasis is somewhat different as large scale map coverage does not exist. Cadastral or urban mapping is usually the responsibility of the local administrative authority, while resource inventory mapping is often the responsibility of state authorities or commercial companies. As a result, photogrammetric data collection has a much more significant role than in the UK. This distinction between the UK and the USA, influenced by the availability of large scale national mapping, is reflected in other countries.

11.4 Recent technical advances

The quality of aerial photography has been improved in recent years with two major developments. The first of these is the forward motion compensation (FMC) aerial survey cameras which compensate for the aircraft's forward motion during the period when the camera shutter is open. These cameras, such as the Carl Zeiss RMK TOP, have greatly reduced the sources of scaling error which occur in aerial photography (see Figure 11.2).

Added to this is the use of gyro-stabilised camera platforms which compensate for the aircraft's roll, pitch and yaw motions, high performance lenses and filters and integrated navigation instruments for automatic navigation and overlap control.

Integrated solutions are being developed with digital aerial survey cameras from which imagery may be directly loaded into image processing or GIS systems. Part of this process is the automatic collection of data related to aircraft position and camera parameters. One such system that acquires and records the airborne data and links to a ground station where the data are integrated with a map database was described by Trone (1991).

The second development has been the introduction of new high resolution film for aerial survey. It is possible to enlarge from the original film up to 15 times without loss of definition and colour quality which means that aerial photography can be used for a greater range of applications (Cassettari, 1992b). As an example one photographic survey could be the source of a regional habitat or land use survey typically at scales ranging from 1:25,000 to 1:10,000 and also provide the source for street furniture inventories from

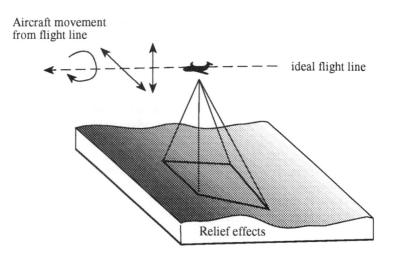

Figure 11.2   Sources of scaling error in aerial photography.

enlargements at a scale of 1:3,000. Figure 11.3 is a black and white reproduction of part of an aerial photograph at the nominal flying scale of 1:25,000 and an enlargement to a nominal scale of 1:3,000, showing the detail that can be extracted.

11.5 Photo interpretation

The alternative to a very precise photogrammetric approach to data collection is to extract information from an aerial photograph by photo interpretation, using simple extraction techniques such as film overlays and pen. Typically photo interpretation will utilise both three dimensional viewing using stereoscopes and mono images. The method leads to the generation of spatial datasets which contain all the errors inherent in the original imagery.

An example of this type of exercise was reported by Young (1992). This described the collection of information for the UK National Rivers Authority, Thames Region, for two rivers, the Lee and the Charwell. This project used aerial photographs to identify land use change that had taken place from the date when the latest maps were published and to add to the base map details relevant to those responsible for managing the river courses. The additional information

Figure 11.3 Example of the enlargement possible from modern aerial survey film without the loss of image quality.

included such data as pollution sites and land use change. A series of 'target notes' or textual descriptions relating to features of interest was also compiled.

The example shown in Figure 11.4 is an extract from a photo interpretation exercise undertaken using 1:10,000 imagery for the purposes of detailing land use change. The importance of this type of data collection is the quantity of notes added to the extraction trace and the detail they convey. Many such exercises involve the creation of hundreds of feature-related notes, the detail from which it is not possible to show using traditional cartographic representations. Such traces are more akin to the model of a descriptive attribute linked to a spatial object or location typical of GIS.

11.6 Orthophotos

An aerial photograph which has been corrected to remove the distortions due to relief and the movement of the aircraft is called an orthophoto. An orthophoto should have the locational accuracy of a map of comparable scale.

The production of orthophoto maps with limited cartographic overlays was used widely by the former UK Directorate of Overseas Surveys (DOS) in Commonwealth countries when producing first time cover of topographic mapping. The concept of generating a geometrically corrected airphoto mosaic for an area defined by map sheet boundaries, adding a limited amount of interpreted information such as main roads, settlements, names and locally significant features like wells, had many benefits. In some cases contours were added to enhance the relief depiction.

The orthophoto map is relatively quick to produce, providing a satisfactory interim solution in areas with no suitable topographic mapping. In some areas these maps have persisted as the only available mapping, while the Seychelles and some Pacific coral atolls have used the orthophoto as the basis of their standard topographic series. The positional accuracy of features depends on the quality and amount of ground control. For many of the less well-developed countries, where the orthophoto was extensively used as the topographic base, positional accuracy was not the over-riding criteria; rather the relative position of features was of importance. However, more rigorous techniques have been used in larger scale orthophoto series, particularly where they form part of the cadastral base mapping. Examples of such large scale orthophoto series include the German 1:5,000 Grundkarte and the South Australian 1:10,000 series.

The concept of using aerial photography to depict the general

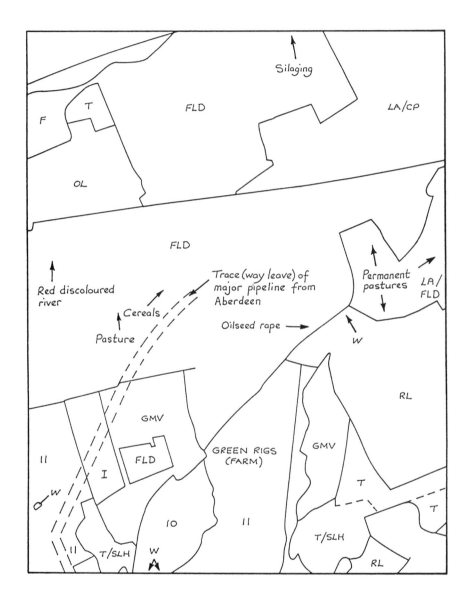

Source: The Landscape Overview

Figure 11.4 Example of a photo interpreter's extraction trace.

terrain and to enhance this with limited cartographic annotation forms the basis of digital image-based systems. They require a different combination of interpretation skills but can provide users, particularly those without the necessary cartographic understanding, with an effective geographical representation.

Significant advances in image display and storage technology have made it possible to use orthophotos within GIS. Such imagery does, however, require much more storage than for equivalent maps and more computer memory for the effective manipulation of the image. As an example of the storage requirements, a 24 bit colour 9 inch aerial photograph scanned at a resolution of 30 microns will require approximately 120 Mb of storage (Cassettari, 1992a). High resolution scanners such as the Zeiss PS1 Photoscan will commonly achieve scan resolutions of 7.5 microns.

The production of digital orthophotos can be achieved using various specialised systems. These require geodetic control and digital elevation data in order to achieve a satisfactory restoration of the original image. The generation of digital orthophotos requires considerable computer processing capability.

Digital orthophotos may be used for the extraction of geo-referenced data directly into vector or raster overlays, with a high order of accuracy (Shanks and Wang, 1991). But when used as a backdrop to map data, discrepancies between the layers may occur. This is due to the cartographic interpretation of features and the rules which govern symbol placement. A road symbol on a map will be wider than the actual road feature and when adjacent to a river will be displaced so that both features can be seen. As a result a map showing roads along a river valley will not overlay exactly on an orthophoto.

11.7 Digital photogrammetry

While this text is not intended to focus on the methods for collecting spatial information it is worth briefly reviewing the developments in digital photogrammetry which are leading to the creation of integrated image-based systems. The traditional analogue machines have been superseded by analytical plotters, and these in turn are being supplemented by digital photogrammetric workstations (Dowman *et al.*, 1992; Welch, 1992).

The principles for photogrammetric data capture remain the same but the solutions currently available both use digital images and undertake all the necessary calibrations to extract data on screen. There are a number of approaches that have been adopted:

- The intermediate point between an analogue and full digital workstation is an analytical photogrammetric plotter. This still uses film diapositives of the aerial photographs housed within the machine together with a mechanical correction of the images to achieve the correct model, but all the adjustments are calculated and carried out internally within the machine and not by the operator. Viewing of the images still relies on optical viewers.
- Low cost digital photogrammetric workstations are becoming available based on PC platforms which use one of two principles. They either display the two overlapping portions of digital aerial photographs on a split screen and view these using simple optical viewers attached to the computer screen, or they use the anaglyph principle to create the three dimensional model where the user views slightly displaced red and green images through glasses. This latter method only operates with black and white photographs.
- High cost, fully digital photogrammetric workstations use high resolution screens and display the finest resolution images. They display the overlapping images together on a single screen, slightly offset but still overlapping. They are viewed by the operator wearing polarised glasses in which the polarity of the two lenses is different and is alternated at a rapid rate. Thus the viewer sees each image in one eye for a fraction of time before seeing the other image in the other eye. In this way the brain is able to construct the three-dimensional model. All the adjustments are undertaken interactively to maintain the model for the operator.

Such solutions can be directly linked into a digital mapping system or a GIS so that photogrammetrically captured data can be fed into geo-referenced databases (for example Brown, 1991; Shanks and Wang, 1991). The cost of such solutions remains high and operators need skills in both the photogrammetric theory and the computing, although the manual operating skills are less important. The danger with the more sophisticated solutions is the ability to capture data that contains errors but is believed to be highly accurate due to lack of awareness of photogrammetric principles.

A further development is the digital aerial survey camera that integrates directly with the photogrammetric workstation. Greene (1992) described a digital image acquisition and processing system that utilises GPS positioning to effect rapid image rectification with a minimum use of control points. This used existing image processing software for photogrammetric mapping projects.

11.8 Seamless aerial photography

The problems of orthophoto generation and digital photogrammetric solutions are the high overhead in computer resources. This restricts their use to specialist applications by virtue of the costs incurred, which include hardware and software, specialist staff training and the time required to extract the high quality data. A significant problem for most users is the storage requirements of large survey areas with ready access by multiple users.

An alternative is the use of lower resolution, partially corrected, digital aerial photography (Cassettari, 1992c; Young, 1992). The considerations for this type of solution are:

- The ability to display high quality images on a low cost platform.
- The ability to store whole survey areas, up to thousands of images, in a readily accessible form.
- The ability to generate the digital image at a cost comparable to the printed image, so that digital solutions are as attractive to the user as traditional printed photographs.

The solution described by Cassettari (1992c) uses a simple linear transformation to effect a partial correction which is acceptable in most areas except those with extreme relief. This correction is undertaken in a production environment by identifying common features on a digital image and a digital map of the same area which are then used to calculate a transformation. While not an orthophoto, this approach is a considerable improvement on the use of uncorrected images which is common with photo interpreters.

The overlap between images in an aerial survey results in much duplicate cover. This was solved by the creation of geo-referenced tiles or edge-matched images identical in concept to the map sheet, as shown in Figure 11.5.

The system described by Cassettari (1992c) runs on a standard Apple Macintosh solution with the only hardware enhancement being a 24-bit graphics board to achieve the best image quality. Each tile fits a computer screen window which is 15cm square, with the image scale being determined by the original scale of the imagery. The image resolution is 72 dpi which gives a good quality image but does not allow any zooming in without the pixel structure becoming apparent. The advantage of using such a coarse resolution is that each tile has an uncompressed storage of about 600K, and compression can achieve a file size of less than 100K. In this way hundreds of tiles may be stored on a single CD-ROM.

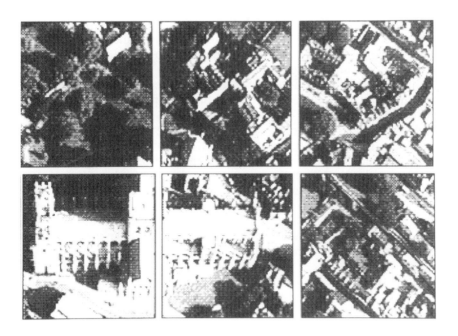

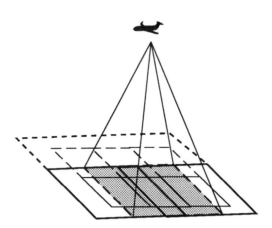

Figure 11.5 Aerial photographic 'tiles' (after Cassettari, 1992c).

The main advantage of such an approach is the availability of a large area of aerial photography managed in the manner of digital map data. Information can be extracted directly into digital map form with the only constraint being the inherent accuracy limitations of the tile. Such an approach has all the apparent advantage of map-based systems plus the additional value of the aerial photograph for additional detail and context.

It is possible to perceive a scenario in the not too distant future where the increase in computer capability coupled with decrease in cost will mean that the orthophoto and digital photogrammetric solution will merge with the low-cost image system described above. The availability of low cost, high resolution orthophoto 'tiles' will have a major impact on the development of image-based systems, and may lead to changes in the way GIS users perceive the need for map data.

While the aerial photography will not supersede the map as a source of geo-referenced information, it should be regarded as a complementary source. It is therefore necessary to look at integrated solutions for information management where the photograph and the map are analysed by the users automatically to generate derived information.

11.9 Changing role of aerial photography in GIS

The previous sections have sought to emphasise the ways in which the use of aerial photographs is changing. For many the idea that the map may be replaced by an image is not consistent with the current perception of a spatial information system. Aerial photographs are generally not as widely used as maps and therefore many potential users are not familiar with their advantages and their limitations.

The aerial photograph will still retain its original role as a source of spatial data either collected to a very high order of accuracy using photogrammetric solutions or to a lesser order of accuracy using lower cost solutions. These processes may be regarded as 'stand alone' with the outputs being passed to other information systems or they may be integrated into the information collection and management process. This will depend on the amount of data collected from aerial photographs and their importance to the organisation.

The aerial photograph also has an increasing role as the geo-referenced base information on which user-specific details are displayed. The image may replace the map or it may be just one layer that the user selects in order to build a representation of the area of interest. The greater availability of orthophotos, or at least partially corrected images, will lead to greater integration of map and photograph.

This greater integration will also lead to new strategies for data collection, where information is extracted from the aerial photographs by tracing features using a mouse or puck. This has implications in terms of the way in which spatial information is managed since the data collection moves from the domain of the photogrammetrist or specialist photo interpreter to the GIS user. Thus the GIS user has to have an awareness of the nature of aerial photographs and how to interpret them.

Aerial photographs could become the base information for many spatial information systems. The most sophisticated image-based systems will display raster images with vectorised map data in a manner which enhances user visualisation of the subject data, will incorporate a sophisticated raster/vector integration strategy within the data handling, and will include image processing as part of the analytical capabilities that it provides. The increased use of aerial photographs in GIS should be viewed as a move towards a greater level of information integration, which has the potential to enhance the use of such data. There are also a number of problems, particularly how to effect successful integration strategies.

11.10 Remote sensing

The alternative to aerial photographs is remotely sensed data. Some photographic images are taken from space, such as those collected from the Russia spacecraft using cameras with a focal length of 36 inches. The exposed film is sent back to Earth and retrieved for processing. Such imagery is of large areas and is suitable for regional or national surveys, has some of the inherent characteristics of aerial photographs, but can not be used for photogrammetric data capture.

The alternative image-based data is that derived from electromagnetic sensors and sent directly from the satellite to a receiving station in digital form. The types of sensors and their attributes are discussed in many volumes, for example Barrett and Curtis (1982), Curran (1985) and Cracknell and Hayes (1991). The most significant systems are summarised in Table 11.1.

A new generation of remote sensing images is planned with higher resolutions, including a 5 metre resolution for the next generation of SPOT satellites. As resolution increases the value of the data potentially increases, with the amount of detail that can be extracted. A further enhancement is provided by the SPOT panchromatic system which flies overlapping swathes, thus giving images that can be used for three dimensional work.

Satellite	Orbit	Application	Resolution	Repeat	Life span
Spot - 2 XS Panchromatic	Sun-synchronous Near polar	Earth Observation Mapping	20 m 10 m	16 days or 5-10 days	Jan 1990 -
Landsat 6 TM	Sun-synchronous Near polar	Earth Observation Multispectral data	30 m 15 m	16 days	July 1993 -
Landsat 5 TM (Thermal Band 6) Multi Spectral Scanner	Sun synchronous Near Polar	Earth Observation Multispectral data	30 m 120 m 80 m	16 days	Mar 1984 -
Landsat 4 TM (Thermal Band 6) Multi Spectral Scanner	Sun synchronous Near Polar	Earth Observation Multispectral data	30 m 120 m 80 m	16 days	July 1982 -
ERS - 1 ASTER Side Aperture Radar	Sun synchronous Near Polar	Earth Observation (Thermal) Earth Observation (Radar)	1 Km 30 m	3, 35, 176 days	July 1992 -
FUYO -1 SWIR Microwave	Sun synchronous Near Polar	Earth Observation Multispectral data	18x24 m 18x18 m	44 days	Feb 1992 -
IRS - 1A	Sun synchronous Polar	Earth Observation (Indian Territory)	36, 72.5 m	22 days	Mar 1988 - 1991
NOAA 9 - AVHRR	Sun synchronous Near Polar	Earth Observation Vegetation and weather studies	1.1 Km	12 hrs	Dec 1984 -
GOES E/W Meteosat Vis Infra Red	Geostationary Geostationary	Weather monitoring Meterological observation	8 Km 2.4 Km 5.0 Km	30 mins 30 mins	Oct 1975 - Nov 1979 -
HCMM Vis Infra Red	Sun synchronous circular	Earth Observation experimental Thermal IR	500 m 600 m	16 days	Apr 1978 - Sept 1980
Nimbus - 7 CZCS	Sun synchronous Near Polar	Oceanography, meterology, pollution monitoring	800 m	6 days	Oct 1978 - Oct 1984

Table 11.1 The main remote sensing systems.

The range of applications to which remote sensing has been put is extensive, as shown by Table 11.2, but in many instances these applications have failed to develop into a broad-based user community. Remote sensing has remained a rather specialised area of spatial information management due in part to the need for a high level of skills for image processing, the cost of the computing resources and the lack of integration with mainstream GIS applications.

11.12 Image analysis

Much of the data collected from remote sensing satellites is in the form of images which contain a large amount of information. The extraction of this information involves the manipulation of the images. Cracknell and Hayes (1991) identify the purposes of image processing as being:

- The extraction of information from the basic image;
- To emphasise or de-emphasise certain aspects of information contained in the image;
- To perform statistical or other analyses to extract non-image information.

Image processing is a branch of information technology which involves a wide variety of techniques. Some are well known as they are used to manipulate other types of image such as contrast stretching to increase image contrast between features in an image as is used in photography.

Other types of image analysis are more complex, such as principal component analysis and Fourier transformations. These involve the statistical analyses of the images and require user understanding of the statistical methodology employed and the limitations on the results that are derived. Detailed reviews on image processing techniques are to be found in many volumes (see Cracknell and Hayes, 1991).

11.12 Role of remote sensing

During the 1980s a debate on the role of remote sensing as a data collection methodology and a discrete suite of analysis tools led to the creation of an independent discipline. Advocates of remote sensing have, however, never managed to raise remote sensing beyond a narrow set of specialist tools within the spatial data handling community. This can be attributed to various limiting constraints:

- The availability of suitable computer platforms and appropriate storage and output devices at an acceptable cost.
- The complexity of image interpretation for generating high quality solutions to spatial problems, which can be attributed in part to the resolution of the images, the quality and availability

Archaeology and anthropology

Cartography

Geology
  surveys
  mineral resources

Land use
  urban land use
  agricultural land use
  soil survey
  health of crops
  soil moisture and evapotranspiration
  yield predictions
  rangelands and wildlife
  forestry - inventory
  forestry, deforestation, acid rain, disease

Civil Engineering
  site studies
  water resources
  transport facilities

Water resources
  surface water, supply, pollution
  underground water
  snow and ice mapping

Coastal studies
  erosion, accretion, bathymetry
  sewage, thermal and chemical
  pollution monitoring

Oceanography
  surface temperature
  geoid
  bottom topography
  winds, waves and currents
  circulation
  mapping of sea ice
  oil pollution monitoring

Meteorology
  weather systems tracking
  weather forecasting
  sounding for atmospheric profiles
  cloud classification

Climatology
  atmospheric minority constituents
  surface albedo
  desertification

Natural Disasters
  floods, earthquakes,
  volcanoes, forest fires,
  subsurface coal fires,
  landslides

Planetary studies

Table 11.2 The uses of remote sensing (after Cracknell and Hayes, 1991).

of control and ground truth data and the need to develop the applicability of analytical techniques to problems.
- The skills base required to utilise image processing systems.
- The lack of integration into broader spatial decision-making strategies.

The complexity of image processing has been one of the main causes for limiting the use of remote sensing to a research environment or to highly specialised commercial applications. Even such techniques as image classification and neighbourhood averaging and filtering are too involved for most non-specialists. One of the greatest impacts of remote sensing and image processing has been in the generation of high impact true-colour images for posters and display purposes.

The areas in which remote sensing and image analysis has had most to contribute until recently were problems on a continental or global scale associated with the increased awareness of environmental issues. The 1990s have seen a resurgence of interest in remote sensing capabilities which reflects the growth in concern over the deterioration in the global environment. Monitoring rain forest removal, ozone depletion, desertification and climatic change are all topics of social and political concern to the international community.

Together with this change in the public perception of important topics, particularly in the richer developed countries, was the reduction in cost and increased capabilities of suitable hardware platforms which has introduced remote sensing to a lower level of user. The wider use of relatively cheap high resolution screens, 24 bit colour and the introduction of mass storage devices will continue this trend. There is, however, still a need to develop image processing software to promote the wider use of the imagery.

The recently developed ability to relate satellite images to topographic map data, ground survey and accurate position fixing using GPS, adds an extra dimension to the use of the images. Examples of the integrated applications are growing. The use of Landsat Thematic Mapper images at 30 metre resolution combined with SPOT 10 metre resolution panchromatic images and overlaid on digital elevation models to generate perspective terrain representations is one such example. The increasing number and range of satellite sources is likely to provide new approaches for combining images, including the integration of multispectral and radar images with digital aerial photographs.

The use of image analysis techniques has not, in general, been integrated within broader image management systems and therefore can at present only play a limited role in spatial image-based systems. In much the same way that specialist GIS analysis is likely to remain within the remit of the GIS professional, even when simpler GIS capabilities are integrated into broader decision-making strategies, so image analysis will be the remit of the remote sensing specialist. There remains, however, the challenge of developing analytical procedures which allow the user to enhance the original image and to develop structured methodologies for integrating the various data sources.

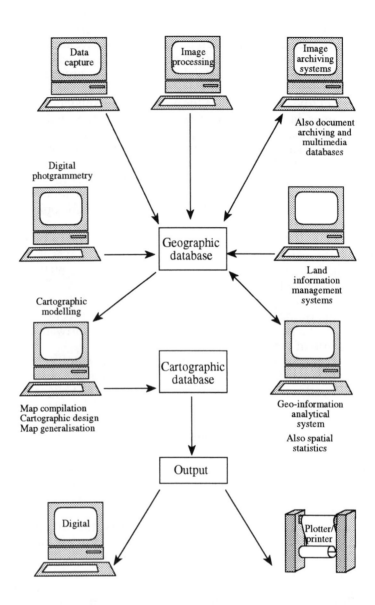

Figure 11.6 Integrated suite of applications software bringing together GIS and image-based solutions for the creation, manipulation and display of spatial information.

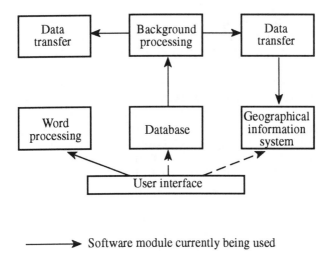

                    Software module currently being used

Figure 11.7 Integrated spatial information system based on a multi-tasking environment.

11.13 Integrated image-based systems

The challenge for those establishing spatial information management strategies is the effective integration of appropriate aerial photographs and remotely sensed images into broader GIS solutions. The important considerations are firstly how to make effective use of these data sources for the collection of interpreted information and secondly how to harness the advantages of visual interpretation and spatial context offered by image-based systems.

While it is possible to conceive of a single GIS toolbox containing map-based, photogrammetric and image analysis functionality the end product would almost certainly be too complex for many GIS users. One solution is to use a software module approach with transparent data transfer from which an integrated solution may be built to meet the needs of individual users (Figure 11.6). Such an approach is adopted by larger software developers, such as Intergraph, wishing to produce products in all the relevant areas.

A simpler solution is the adoption of a data transfer standard by two or more vendors that allow data to be imported and exported across specified packages in a transparent way for the user. The combination of ARC/INFO and the ERDAS remote sensing package is an example of such an approach. The difficulties with such solutions are the commitment of the user to a single or limited number of vendors and the less than transparent integration of the data types from the user's perspective.

The establishment of widely accepted data transfer standards will provide the basis on which system vendors can develop integrated solutions of this sought. As previously discussed, the crux of integrated geo-information management lies in the user's ability to move interpreted data between spatial and non-spatial systems.

A more beneficial approach is that offered by an open systems environment, where multi-tasking can be operated (see Figure 11.7). The use of WIMP type interfaces facilitates the use of more than one package with the display of maps, photographs and remotely sensed images in independent package windows. This gives the user the advantages of map-based GIS analysis adjacent to the display of image-based data. The flexibility of this type of solution is that the user chooses the data to be displayed and decides how it is best presented, with the ability to have many different data types displayed all at the same time.

This is of course not without its problems. The quality of the user interface, understanding the issues of data accuracy, quality and currency, as well as the technical issues, can all affect the use of such an approach.

References

Barrett, E. C. and Curtis, L. F. (1982) *Introduction to Environmental Remote Sensing.* Chapman and Hall, New York.

Brown, R. O. (1991) Photogrammetric GIS technology: Feature mapping on digital stereo imagery. *Proceedings GIS/LIS,* Atlanta, pp. 882-90 .

Cassettari, S. (1992a) Towards integrated image-based systems. In: *Geographic Information. The Yearbook of the Association for Geographic Information.* Eds. Cadoux-Hudson, J. and Heywood, I., Taylor and Francis, London, pp. 201-6.

Cassettari, S. (1992b) The development of air photo image-based systems: The UK perspective. *Proceedings GIS/LIS,* California pp. 87-94.

Cassettari, S. (1992c) Geo-referenced image-based systems for urban information management. *Proceedings 15th European Urban Data Management Symposium,* Lyon, pp. 463-70.

Cracknell, A. P. and Hayes, L. B. W. (1991) *Introduction to Remote*

Sensing. Taylor and Francis, London.

Curran, P. J. (1985) *Principles of Remote Sensing.* Longman, London.

Dowman, I. J., Ebner, H. and Heipke, C. (1992) Overview of European developments in digital photogrammetric workstations. *Photogrammetric Engineering and Remote Sensing,* Vol. 58, No. 1, pp. 51-6.

Greene, R. H. (1992) Digital photogrammetric mapping integrating GPS in acquisition and GIS processing. *Proceedings GIS/LIS,* San Jose, pp. 293-99.

Kirby, R. P. (1992) The 1987-89 Scottish national aerial photographic initiative. *Photogrammetric Record,* Vol. 14, No. 80, pp. 185-200.

Shanks, R. and Wang, S. (1991) Integrating digital orthophotography and GIS: A software-based approach to vector/raster processing. *Proceedings GIS/LIS,* Atlanta, pp. 674-82.

Trone, D. L. (1991) Integrating flight and photographic data into the GIS/map databases. *Proceedings GIS/LIS,* Atlanta, pp. 828-33.

Welch, R. (1992) Photogrammetry in transition - analytical to digital. *Geodetical Info,* Vol. 6, No. 7, pp. 39-41.

Young, R. N. (1992) Digital imaging systems link airphoto interpretation to GIS. *GIS Europe,* Vol. 1, No. 3, pp. 34-7.

12

Multimedia and hypermaps

12.1 The multimedia concept

As previously discussed, the power of spatial information technology for improving decision support is through the enhancement of the map and attribute data by spatial analysis techniques and the integration of map data with other data types. Coupling various analysis methodologies with the powerful retrieval and display components provides for a much more comprehensive analytical tool.

The view persists that the GIS is essentially a map-based tool to which may be added non-spatial data, even where the the non-spatial data constitutes the greater proportion of the dataset. The alternatives were discussed in the previous chapter about image-based, rather than map-based, systems. It is necessary to extend this further and address other types of data and methods by which these may be handled in a spatial information handling environment.

The vertical, survey standard aerial photograph and satellite images form two groups of data types, but there are a wide range of others, which are either inherently spatial in nature or may be used to supplement or complement spatial data. These include aerial oblique photographs, ordinary hand-held or still photography, film or video images, graphics, graphs and statistical data, animations and cartoons, sound, continuously recorded data such as water quality monitors or seismic recorders. This broad variety of data means is generically referred to as multimedia.

The traditional view of a spatial database may now be extended to encompass digital forms of many of the above data types. This has led to new integrated spatial databases in which the geo-reference is retained as a key organisational element but in which the map does not necessarily play a part. This in turn has led to the definition of new means by which the data can be accessed, especially the user query environment.

This chapter considers the types of multimedia data and how these may be used in a GIS and goes on to consider the concepts of hyperdocuments and hypermaps in a spatially organised information handling environment.

12.2 Multimedia data

Multimedia may include any type of data collected using a variety of methods. These include imaging devices such as still camera, video camera or sensors, sound recording devices, data loggers and computer-based text and graphic databases. They characteristically collect the data in a digital form or the data may be converted into a digital data.

For the purposes of spatial information systems, these data may be regarded as either inherently spatial in character or not, depending on the subject as well as the media used. For example a hand held photograph of a person has no inherent spatial properties even though it may be possible to identify the location of that person. Of more importance is the individual and the reason for the taking the photograph such as a holiday snap or fashion image. The date is probably a more critical reference. On the other hand a still photograph of an archaeological dig may contain basic spatial context of features such as the settlement pattern and the position of unearthed pottery fragments. These may be important in the later analysis of the findings.

There have been a number of attempts to bring together these various forms of data. The British Broadcasting Corporation (BBC) were responsible for the Domesday Project in the mid 1980s (Openshaw and Mounsey, 1986; BBC Enterprises, 1986). This was a collection of still images, maps, tables of numerical data and text information about places across Britain compiled to commemorate the original survey of 1086 for William the Conqueror. The information was made available in the form of two videodisks and contained a number of different strategies for accessing the information, including spatial location.

Another example is the videodisk created by the National Geographic Society which combined maps, pictures and film to create a resource for teaching history and geography. The following sections consider some of the multimedia data types which have not been addressed elsewhere in more detail.

12.3 Digital video

Video film taken using widely available recorders may now be converted into a digital form for use in multimedia systems. Each frame used in the digital film is converted into a digital form and stored as a still image. The type of system and the speed and memory of the computer dictates how many frames are used to reproduce the film. Most smaller PC-based systems are currently not able to reproduce the 25 frames a second used in standard video film. The images are typically stored in a compressed format and uncompressed as required in replaying the video. Each frame may be treated as if were a still image and extracted from the video sequence for use in other ways.

Video has been taken for the purposes of extracting feature location, for example a study is being undertaken in the UK for British Rail which wishes to establish a track side furniture inventory. A video is taken from the cab of a train running along the section of track under consideration. At the same time distance along the track is measured and the two data sets are tied together by the use of synchronised time recording. Each video frame has superimposed on it a bar which is calibrated at the start of the run so that when a piece of trackside furniture is seen by the analyst to be on the bar an offset from the track and distance from start of the run may be calculated. This type of data can be used to give relative positions quite easily and by relating this to a digital map of the precise track alignment an accurate geo-referenced location.

Video has also been used as the source of still images for inventory purposes. A Scottish company have produced a digital database of all commercial properties in the country. This includes ownership details, locations shown on a map and pictures of the property extracted from video images.

12.4 Sound

The notion of linking sound to a map extends beyond that of a multimedia linkage. Sound can be associated with spatial objects as an attribute, such as the sound of church bells associated with the spatial representation of the church (Figure 12.1). The attribute is not quite like that normally associated with a spatial entity in that playing the sound involves at least a consideration of the sound's duration and volume. Furthermore, the sound has no permanent or semi-permanent existence in the way a table of attributes has and may have to be repeatedly played to be understood.

Sound also has spatial characteristics. A sound emitted from a source at a particular time will fill a definable space after a given amount of time. The characteristics of that sound and how it changes across space can be modelled as if it were a special type of spatial object. In its most complex form the analysis of sound in a GIS can be regarded as a multi-dimensional information problem - that of $x,y,z,$ time and the sound itself (Cassettari and Parsons, 1993).

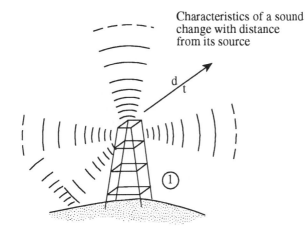

Characteristics of a sound change with distance from its source

d/t

— — — Limit of audible sound, which may be used to mark the spatial extent of a three dimensional sound object

d/t   The nature of a sound object changes through time and with distance

①&②   Multiple spatial objects are formed as a consequence of reflected sounds. The result is complex four dimensional spatial entities with attributes that vary in a predictable manner across the object

Figure 12.1 Simple data model for adding sound to a spatial information system.

Information about a sound or overall measures of sound and their relationship to the environment are typically presented cartographically by the use of contours which describe a surface. This surface is essentially the interface between the sound object and the terrain that represents a measure of sound at the ground surface. These types of maps are prepared for airports where sound footprints are defined for various aircraft types and their flight paths. Noise pollution can be limited by modelling the advantages and disadvantages of changing the steepness of the climb after takeoff or the orientation of the flight path.

In many ways this two dimensional cartographic representation of sound is inadequate. There needs to be a more integrated model for handling the audio aspects of data and for undertaking integrated analytical procedures. Cassettari and Parsons (1993) discussed the need for a true four dimensional data structure in which the characteristics of a sound object change within that object in a predictable manner. Thus a sound object will have a value which is louder nearer to the source and quieter further away from the source. The change in these values can be calculated on the basis of the properties that affect the propagation of sound. This type of complex modelling is required to achieve a full integration between sound and other spatial data types. It is, however, only likely to be developed for very specialised applications such as the prediction of noise pollution around airports.

12.5 Animation

The use of animation is widespread in the communications, media and education fields. There are well-developed processes for creating animated sequences and it is now common to see animations integrated with video sequences. This has been done for full length feature films and is used widely for television advertising.

Animations may be used to present information on spatial process. Animated sequences showing the effects of particular meteorological events or geological processes have been used in education and very simplified animations of weather sequences are regularly used in television weather forecasting. These types of animation can be used where it is not possible to obtain real film or the detail on real images is simplified to facilitate ease of understanding.

The use of animation can be extended into a spatial multimedia context. Animations may be used as a supporting or complementary data type which adds to the understanding or communication processes, or they may be integrated with other spatial data types for the purposes of predicting or visualising spatial relationships that might occur in the future.

Digital animations of buildings are often created for large scale developments in which computer animations are drawn and recorded, frame by frame, onto video. With this technique it is possible to simulate a walk around or a flight over a building. The main limitation with this process is the ability to design a realistic environment, for example the problems of visualising objects like grass, trees and water.

An alternative approach is that described by Ertl *et al.* (1992) which combines video film, computer animation and GIS data for the visualisation of proposed buildings. The objective was to show a new building in-situ as it would be seen from static locations and from moving vehicles. Video footage of the site is merged with a computer generated animation of the building by computing the exact position, viewing direction and zoom adjustment for every frame of the video (Figure 12.2). The animation is based on the architect's design and perspective images are calculated to fit each video frame. Points of known spatial location, 'passpoints', are used to link video image and animation together. The system software, move-X, used several hundred surveyed control points so that the CAD drawing could be controlled in terms of its position and perspective view on each frame.

This process is currently expensive. A considerable amount of processing time is required to generate the CAD image for insertion into each frame and there is a need for at least five accurately surveyed passpoints to be identified in each frame. Its advantages lie in the ability to view development projects where neighbourhood is of critical importance and to evaluate different variants of a planned building.

12.6 Hypertext concept

The idea of developing a form of multi-dimensional information system that contained links between a wide range of data items stems from work by Vannevar Bush in the 1940s when he proposed the 'MEMEX' idea (Bush, 1945). This was developed by Ted Nelson in the 1960s (Nelson, 1965; Nelson, 1967) when he defined the hypertext concept of an undifferentiated link structure between nodes of information as a combination of natural language text with the computer's capacity for interactive branching, or dynamic display of a nonlinear text which can not be printed conveniently on a conventional page.

Essentially the hypertext model can be described as 'a set of nodes connected by undifferentiated links, where the nodes can be abstractions made up from any kind of text or graphic information elements' (Raper et al., 1990). The issues associated with the design of hypertext information systems are the content and organisation of the individual nodes which present a problem of management within the computer environment due to their complex nature and the method used to interact with the user (McAleese, 1989).

The nodes and the associations between them, called links, form semantic units. Each semantic unit may express a single idea or simple data element such as a word or name, but may be much more complex, such as a map or table of figures. Hypertext usually consists of many simple semantic units connected by means of reference links (Laurini and Thompson, 1992). These reference links can take various forms but

Photos by courtesy of Grintec Graz, Austria

Figure 12.2 Frames from a video film with computer animation of a
proposed building.

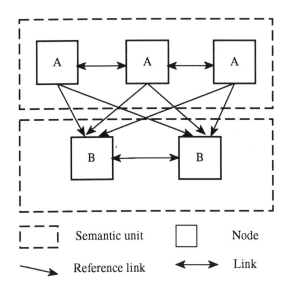

Figure 12.3 Hypertext model of semantic units, nodes, links and reference links.

essentially tie together the various semantic units and the references by which networks may be constructed (Figure 12.3). As a consequence there is the ability to 'navigate' around the data.

The theoretical structures based on Ted Nelson's ideas developed rapidly in the 1970s and 1980s with the introduction of new high level computer languages and the growth in hardware capabilities since hypertext solutions require considerable computer memory.

The release of HyperCard by the Apple Corporation in 1987 made user friendly systems widely available on a personal computer platform. The system's success was based on the ease with which 'stacks' of information can be developed and links made between elements using the advanced object-oriented programming language, HyperTalk (Raveneau *et al.*, 1991). The availability of HyperCard stacks is increased by the software being distributed free to each buyer of a Macintosh computer.

12.7 Hyperdocuments

Most conventional documents are printed text media and have a logical and a physical structure which is essentially sequential and hierarchical (Laurini and Thompson, 1992). The structure of a book is a good example with its various chapters and sections, contents page, bibliography and index. Hyperdocuments on the other hand embody the hypertext concept in that they are essentially organised in a non-linear way and can make use of various multimedia sources.

A non-digital form of the hyperdocument is the atlas. It contains various forms of information including maps, photographs, text, tables and diagrams which are linked through an index and a spatial referencing structure. The user does not use an atlas like a novel, starting at page one and progressing to the conclusion, but instead uses the index to find maps and related information about an area of interest. The chances are that the various pages are to be found in various locations within the atlas.

The idea of hyperdocuments was used in the development of a HyperCard tutor for GIS (Raper and Green, 1989; Raper et al., 1990). This uses a number of strategies whereby the user can move around the available information. The simplest form is a sequential progression through the tutor but there are also loops which develop a particular theme, the opportunity to divert to another part of the tutorial if a particular word or subject is not understood, and the chance to miss out sections on a particular subject. Figure 12.4 demonstrates this model.

There are an increasing number of examples of hypertext systems which are being used to integrate maps and images with other data types. The 'Great Cities of Europe' project is a multimedia database for use by architects and urban planners. The objective is to provide information in the form of maps, plans, slides, video and sound, about the structure, developments, problems and planning policies in a sample number of European cities (Polydorides, 1992). The objective is to use hypertext techniques to create a database which can be used to describe how a city grew, its present structure, urban problems and the measures implemented or planned for urban renewal. It is essentially a strategic overview of a city, in which people with different backgrounds and interests can 'navigate' about the multimedia database to build up their own information subsets on the issues of urban development.

12.8 Hypertext user interfaces

One particular application of the hypertext environment has been the development of more intuitive user interfaces for GIS packages. The traditional GIS package was command driven and only in the last couple of years have WIMP (windows, icons, mouse, pointers) interfaces been developed. Characteristically the user who is very familiar with a

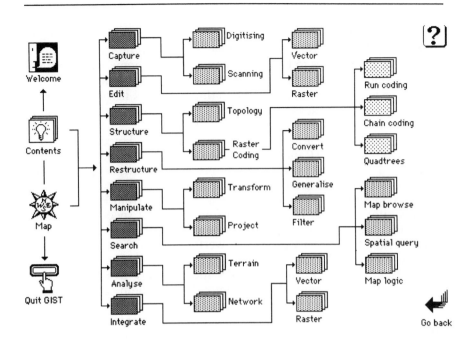

Figure 12.4 GIS Tutor hyperdocument model (after Raper *et al.*, 1990).

package and the process to be carried out will often prefer a command interface as the quickest. Conversely the user with little knowledge of a package or the process to be undertaken is more at home with a WIMP interface since it is possible to experiment more freely with the package. This latter has implications for generating solutions in which the user has a complete understanding of the operations used and the consequences to the data.

An example of a linkage between a GIS and HyperCard is the HyperARC system (Raper *et al.*, 1990; Raper, 1991). This uses a HyperCard front end which uses icons and buttons to replace the ARC/INFO command driven interface current at the time (Figure 12.5). It is also possible for the user to interface directly from HyperARC to the GIS tutor so that it is possible to find out about specific processes before they are run. This linkage of a powerful GIS, intuitive front end and hypertext help facility offers a powerful model for the development of more user friendly GIS solutions. The objective of integrated geo-information systems is to make GIS more widely available without over-simplifying its capabilities. Comprehensive help facilities using hypertext techniques offer one way round the problem.

A further study into the development of this type of interface was the UGIX project (Raper and Bundock, 1991). This developed the metaphor used to make the computer interface more intuitive. The Apple Corporation metaphor of a desktop with features such as folders and a dustbin has been very successful in opening the use of personal computers to a wider audience.

12.9 Hypermaps

Hypermaps are an extension of the hyperdocument concept in which spatial referencing is added to the multimedia hyperdocuments. There are two aspects to the spatial referencing of hyperdocuments (Laurini and Thompson, 1992):

- Spatial referencing of document nodes. Since a node represents either a single or a few semantic units it may have one or many spatial references.
- Spatial referencing of maps or other cartographic documents. Each of these documents may have many nodes.

The spatial data model used for hypermaps needs to be considered. The important aspects are the document-to-map relationships and the map-to-map relationships. Laurini and Thompson (1992) describe how this may be achieved using Peano relationships or R-trees and map

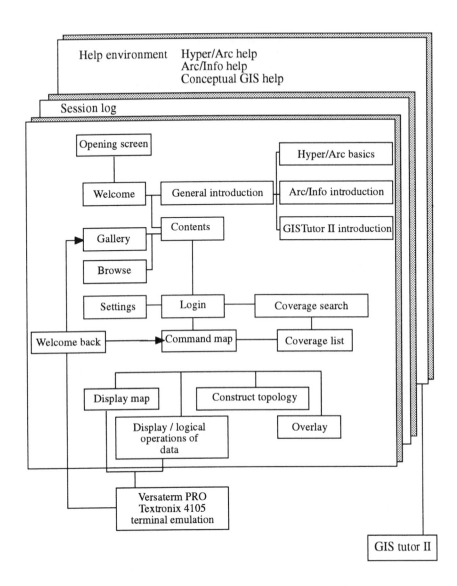

Figure 12.5 User interface from the HyperARC system

pyramids. The important problem to be overcome is the need to differentiate between thematic navigation such as is normally undertaken in a hypertext environment and the spatial navigation which is particular to maps.

The thematic navigation allows the user to move from one node or semantic unit to any other. Different levels of navigation complexity may be implemented, depending on the abilities of the user.

Spatial navigation moves the user from a spatial location to a node or semantic unit. In many GIS solutions this might be referred to as the point-in-polygon or line-in-polygon type query. A single map object or a group of map objects are identified on the basis of a spatial search. The search limits are defined by a user-drawn area or by an automatically created area such as a 'zone of influence'. The subset of map objects is identified in terms of their nodes, semantic units and links, thus allowing the user to navigate from the map to other hyperdocuments. A model of the spatial navigation issues in a hypermap environment are shown in Figure 12.6.

12.10 Cartographic representation in multimedia

It is clear from the discussion in Chapter 10 that a digital environment offers alternative visualisation methodologies for cartographic representation of spatial data. The implications for communication and visualization are further extended when the map forms part of a multimedia database.

For the cartographer it may be desirable to integrate these various data types in a digital environment for the purposes of enhancing the cartographic image by providing information which would not otherwise be included on a map. The importance of the link between GIS and multimedia is the establishment of a wider geographic context in the mind of the user. This is an aim of the 'Great Cities of Europe' Project mentioned in section 12.6 and is important in compiling tourist maps or educational tools.

Parsons (1982) demonstrated the use of maps and aerial photographs in a prototype information system for Covent Garden, London. This uses an aerial photograph and map base, between which the user can alternate to aid interpretation. These act as an index to information about points of interest. A variety of sound and picture icons identify features on the map and act as visual and audio cues to the types of multimedia data available. A video of a particular area is identified by the search icon changing to a representation of a television accompanied by the sound 'Action!'.

The potential for using hypertext and HyperCard for cartographic communication was shown by Raveneau et al. (1991) in the development of electronic atlases. These were Apple Macintosh-based atlases covering the North American French-speaking communities and

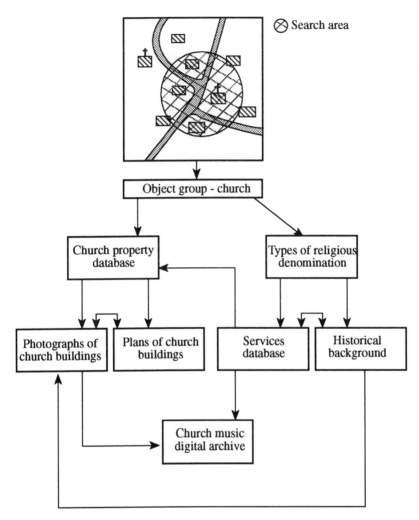

Figure 12.6 Spatial navigation in a hypermap environment.

mines and minerals. The importance of the hypertext concept was in the ability to make 'multiple associations between different elements of information' and 'the possibility of representing and visualising information by various graphic means' (Raveneau *et al.*, 1991, p205).

The importance of both these approaches was that for the user to browse freely but coherently, an appropriate navigation structure has to be created. Also important is the screen format which has to contain map and graphic images plus all the necessary navigation tools (see Figure 12.7). Mechanisms exist which indicate that change in images is occurring. Parsons used an interactively changing icon associated with sound while the atlases use five different visual effects (dissolve, chequerboard, scroll, wipe and barndoor) to indicate transition between different parts of the atlas.

One of the limitations of using Hypercard for this type of cartographic representation is the small size of the normal Macintosh computer screen. Inevitably the small map window means that either the image is small and cluttered or only a small portion of the map is visible at one time. This was a particular problem in the development of the image-based system described in Chapter 11 (Cassettari, 1992) and led to the development of a second version which could have resizeable windows for use on large screens.

This requires the development of alternative cartographic strategies for small maps to ensure effective communication. In particular it is important to fully utilise the interactive nature of the computer environment to enhance the cartographic representation. For example it is possible to have symbols which change colour, flash on and off or change in size. Different symbol sets can be placed on different map layers with only one layer active, while other layers may be changed from colour to grey scale or turned off. The development of these types of 'interactive cartography' require further consideration but are potentially very powerful communication strategies.

12.11 Integrated geo-based multimedia solutions

Finally, it is worth reflecting on the role of multimedia databases as part of an integrated geo-information strategy. One of the major limitations of multimedia data is the storage volumes required. Raveneau *et al.* (1991) notes that the 26 seconds of music at the beginning of their database occupies 290K of memory. The 24-bit colour images described by Cassettari (1992) are composed of 432 by 432 pixels (186,624 pixels) and require about 610K of storage, whereas a 35mm slide stored on a photo-CD can, at the highest resolution, require over 4mb.

Parsons (1992) comments that to some extent these problems have been reduced by the availability of data compression techniques such as the Joint Photographic Experts Group (JPEG) algorithms for images and the various digital video compression algorithms introduced by Apple Quicktime technology. JPEG algorithms may produce compression ratios of between 20:1 and 10:1 with little image degradation (Lewis and Rhind, 1991b). The effect of using such

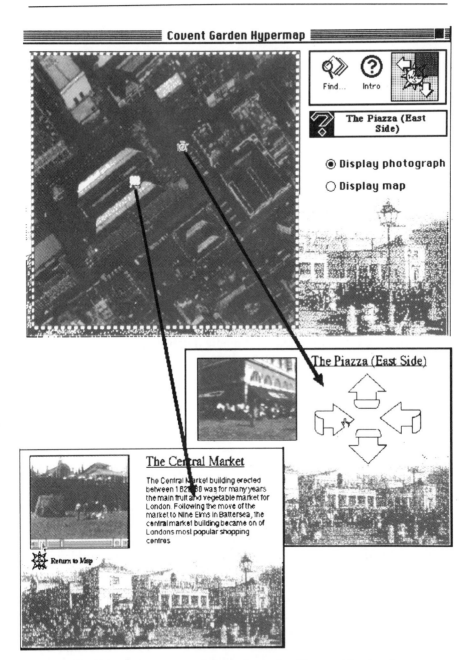

Figure 12.7 Sample screen layout from the Covent Garden hypermap demonstration (Parsons, 1992).

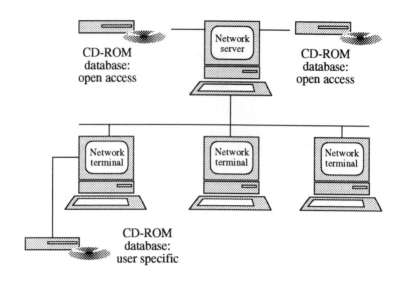

Figure 12.8 A distributed multimedia database using CD-ROM technology.

compression techniques is less need for specialised hardware and storage.

Even so most multimedia databases are so extensive that they require CD-ROM storage. However, the cost of CD technology, including players linked externally to computers or CD players which are integral within a hardware solution, are now common and the cost of mastering and reproducing the CDs is relatively cheap. Thus it is possible to conceive of multimedia data being made widely available within many organisations not on a network but through copies of a CD. Thus the multimedia database rather than being centrally organised can be distributed without the need to consider the problems of networking and communications (Figure 12.8).

The cost of CD-ROMs, while relatively cheap, can be considered significant where many copies are required, particularly if the data are purchased from a commercial source. In the UK costs of CDs range from £30 to £3000. At the higher end of the range multiple copies become too expensive to consider for a large set of users within one organisation. Thus, expensive CD databases are often only available through libraries or central services. The likelihood is that, with increase in availability of CDs and players, unit costs will fall dramatically.

A more important constraint on the CD-ROM is that it can not be updated. Thus a CD-based atlas is as dateable as a printed atlas. Putting map data on CDs requires careful consideration of the currency and revision strategy which an organisation operates. This may be in part alleviated by the move to re-recordable CDs, which will provide mechanisms for transfering data to off-line sites on a regular basis. This will not necessarily replace the on-line database in which the user has immediate access to the most current information.

The integration of multimedia data and hypertext strategies also require consideration in terms of communication and visualisation, access and operating software, user interfaces and data formats. All these need to be addressed if an effective integration is to be achieved successfully.

References

BBC Enterprises (1986) *The Domesday Project* BBC Enterprises Limited, London

Bush, V. (1945) As we may think. *Atlantic Monthly,* July 1945, pp. 101-8.

Cassettari, S. (1992) Geo-referenced image based systems for urban information management. *Proceedings 15th European Urban Data Management Symposium*, Lyon, pp. 463-70.

Cassettari, S. and Parsons E. (1993) Sound as a spatial object. *Proceedings European Conference on Geographical Information Systems,* Genoa, pp. 367-78.

Ertl, G., Gleixner, G. and Ranzinger, M. (1992) Move-X: A system for combining video films, computer animations and GIS data. *Proceedings 15th European Urban Data Management Symposium,* Vol 1, pp. 247-54.

Laurini, R. and Thompson, D. (1992) *Fundamentals of Spatial Information Systems.* Academic Press, London.

Lewis, S. and Rhind, D. (1991a) Multimedia geographical information systems. *Mapping Awareness,* Vol 5, No 6, pp. 43-9.

Lewis, S. and Rhind, D. (1991b) Multimedia geographical information systems. *Proceedings Mapping Awareness Conference,* London, pp. 311-22.

McAleese, R. (1989) Navigation and browsing in Hypertext. In: *Hypertext: Theory into practice.* Ed. McAleese, R., Intellect, London

Nelson, T. H. (1965) A file structure for the complex, the changing and the indeterminate. *Proceedings 20th National ACM Conference*, pp. 84-100.

Nelson, T. H. (1967) *Getting In and Out of Our System, Information Retrieval : A Critical Review.* Thompson Books, Washington.

Openshaw, S. and Mounsey, H. (1986) Geographic information systems and the BBC's Domesday interactive videodisk. *Proceedings Auto Carto London,* pp. 539-46.

Parsons, E. (1992) The development of a multimedia hypermap. *Proceedings Association for Geographic Information Conference,* Birmingham, pp. 2.24.1-6.

Polydorides, N. D. (1992) Great cities of Europe. A multimedia database. *Proceedings 15th European Urban Data Management Symposium,* Lyon, Vol 1, pp. 255-63.

Raper, J. F. (1991) Spatial data exploration using Hypertext techniques. *Proceedings European Conference on Geographical Information Systems,* Brussels, pp. 920-8.

Raper, J. F. and Bundock, M. (1991) UGIX: a GIS-independent user interface environment. *Proceedings Autocart 10,* Baltimore.

Raper, J. F. and Green, N. P. A. (1989) Development of a hypertext based tutor for geographical information systems. *British Journal of Education Technology,* Vol 3, pp. 164-72.

Raper, J. F., Lindsey, T. K. and Connolly, T. (1990) UGIX - A spatial language interface for GIS: concept and reality. *Proceedings European Conference on Geographical Information Systems,* Amsterdam, pp. 876-82.

Raveneau, J., Miller, M., Brousseau, Y. and Dufour, C. (1991) Microatlases and the diffusion of geographic information: An experiment with HyperCard. In: *Geographic Information Systems. The microcomputer and modern cartography,* Ed. Taylor, D. R. F., Pergamon Press, Oxford pp. 201-23.

Appendix A: Acronyms

AA	Automobile Association
AGI	Association for Geographic Information
API	Application Program Interface
ASRP	ARC Standard Raster Product
ATKIS	Authoritative Topographic Kartographic Information System
AVS	Amtliche Vermessungs Schnittstelle
BBC	British Broadcasting Corporation
BIL	Band Interlaced by Line
BIP	Band Interlaced by Pixel
BSI	British Standards Institution
BSQ	Band Sequence
CAD	Computer Aided Design
CCSM	Canadian Council on Surveying and Mapping
CD-ROM	Compact Disc - Read only memory
CEN	Comité Européen de Normalisation
CGM	Computer Graphics Metafile
CIS	Commonwealth of Independent States
CNIG	Conseil National de l'Information Géographique
DBMS	Database Management System
DCE	Distributed Computing Environment
DCW	Digital Chart of the World
DDBMS	Distributed Database Management System
DEM	Digital Elevation Model
DGIWG	Digital Geographic Information Working Group
DIGEST	Digital Geographic Information Exchange Standard
DIME	Dual Independent Map Encoding
DLG	Digital Line Graph
DMATC	Defense Mapping Agency Topographic Center
DOS	Directorate of Overseas Surveys
DSS	Decision Support Systems
DTM	Digital Terrain Model
DXF	Drawing Exchange Format
ECC	Extended Colour Coding
EEI	External Environment Interface
ESS	Executive Support Systems

ETF	European Transfer Format
FACC	Feature and Attribute Coding Catalogue
FMC	Forward Motion Compensation
GADS	Geodata Analysis and Display System
GBF	Geographic Base File
GDA	Geographic Document Architecture
GIMMS	Geographic Information Manipulation and Mapping System
GIS	Geographical Information Systems
GKS	Graphics Kernel System
GPS	Global Positioning Satellites
GSGS	Geographic Section General Staff
GUI	Graphical User Interface
IGES	Initial Graphics Exchange Standard
IHO	International Hydrographic Organisation
ISO	International Standards Organisation
IT	Information Technology
JPEG	Joint Photographic Experts Group
LIS	Land Information System
MACDIF	Map and Chart Data Interface Format
MAUP	Modifiable Areal Unit Problem
MIPS	Millions of instructions per second
MIS	Management Information Systems
NATO	North Atlantic Treaty Organisation
NCGIA	National Center for Geographic Information and Analysis
NES	National Exchange Standard
NIST	US National Institute of Standards and Technology
NTF	National Transfer Format
ONC	Operational Navigational Charts
OS	Ordnance Survey
OSE	Open Systems Environment
OSF	Open Systems Foundation
OSI	Open Systems Interconnection
PAF	Postal Address File

PC	Personal Computer
RDBMS	Relational Database Management Systems
SDSS	Spatial Decision Support Systems
SDTS	Spatial Transfer Data Standard
SERTM	South East Regional Transport Model
SIF	Standard Interchange Format
SLF	Standard Linear Format
SPDFDM	Standard Procedure and Data Format for Digital Mapping
SQL	Standard Query Language
SVF	Single Variable File
TIGER	Topologically Integrated Geocoding and Referencing
USL	Unix Systems Laboratories
VPF	Vector Product Format
VRF	Vector Relational Format
WGS	World Geodetic System
WIMP	Window–Icon–Mouse–Pop-up menu

Appendix B: A summary of the main GIS and related software that has been produced for microcomputers

Package name	Platform		Generic Category						
	PC	MAC	GIS	Carto	CAD	RS	Draw	Digit	Cont
3DTM	X		X						
4BASE/4VIEW	X							X	
AGIS	X		X						
ALDUS FREEHAND		X					X		
APPLETIPS		X				X			
ARCCAD	X				X				
ARC/INFO	X		X						
ARCVIEW	X		X						
ASG TOPO/COGO	X				X			X	
ATLAS AMP.	X			X					
ATLAS*GRAPHICS	X			X					
AUTOCAD	X				X				
AUTOMAPS	X		X						
AUTOSKETCH	X				X				
AZIMUTH		X		X					
BYERS VIEW STATION	X			X					
CAD OVERLAY GIS	X		X						
CADCORE/TRACER	X				X				
CANVAS		X					X		
CARS	X		X						
CARTOCAD	X			X					
CGIPS	X					X			
CLARISWORKS		X					X		
COMPUGRID	X		X						
CONTOUR	X								X
CONTOUR 81		X							X
CRICKETGRAPH		X					X		
DATAMAP LOCATOR	X		X						
DESKDRAW		X			X				
DIAGRAPH	X			X					
DIDS	X			X					
DIGIT-II	X							X	
DOCUMAP	X			X					
DRAFTSMAN	X						X		
DRAGON	X					X			
DRAWING TABLE		X					X		
DREAMS		X					X		

Package name	Platform		Generic Category						
	PC	MAC	GIS	Carto	CAD	RS	Draw	Digit	Cont
EASYCAD	X				X				
ERDAS	X					X			
ESLMap	X						X		
EXECUVISION	X			X					
FARMGIS	X		X						
FASTCAD	X				X				
FREEHAND		X					X		
FREELANCE	X				X				
FULLPAINT		X					X		
GENERIC CADD	X				X				
GEOQUERY		X			X				
GEO/SQL	X			X					
GEOSYS	X			X					
GEOVIEW	X			X					
GGP	X		X						
GIMMS	X		X						
GISPLUS	X		X						
GRASS	X		X						
GSMAP	X				X				
GWN-DTM		X						X	X
HYPERATLAS		X		X					
IDRISI	X		X						
ILLUSTRATOR		X					X		
IN-CAD	X			X					
INFOMAP	X			X					
IRIS GIS	X		X						
LANDTRAK	X		X						
MACATLAS		X		X					
MACBRAVO		X						X	
MACCHORO II		X		X					
MACCONTOUR		X							X
MACDRAFT		X			X				
MACDRAW		X					X		
MACGEOS		X		X					
MACGIS		X	X						
MACGRASS		X	X						
MACMAP		X	X						
MACPAINT		X					X		
MAP	X			X					
MAP AND IPS	X		X			X			
MAPBOX	X		X						
MAPCON		X	X						
MAP-II		X				X			
MAPGRAFIX		X	X						
MAPINFO	X	X	X						

Package name	Platform		Generic Category						
	PC	MAC	GIS	Carto	CAD	RS	Draw	Digit	Cont
MAP-MASTER	X			X					
MAPIT	X			X					
M.A.P.S.	X			X					
MAP STATION	X		X						
MAPVIEWER	X			X					
McMAP	X			X					
MICROGIS	X		X						
MICROMAP	X	X		X	X			X	
MICROPIPS	X					X			
MICROSTATION	X		X						
MIMS	X		X						
MODULAR GIS	X		X		X	X			
MORPHOTERRA		X							X
NUCOR	X				X				
OSU MAP		X					X		
OZGIS	X		X						
PANAMAP	X			X					
PANACEA		X		X					
PC-DRAW	X						X		
PCIPS	X					X			
PCMAP	X		X						
PC MAPICS	X			X					X
PIXEL PAINT		X					X		
PIXEL TRACK	X							X	
PMAP	X		X						
POLYMAPS		X	X						
POWERDRAW		X			X				
PRODESIGN II	X				X				
QUADRANT	X		X						
QUICKMAP	X		X						
RANDMAP		X					X		
ROOTS	X							X	
SLICID	X				X				
SPANS	X		X						
SPASE	X		X						
SPSS GRAPHICS	X						X		
STATMAP		X		X					
STRIPES	X		X						
SUPER PAINT	X						X		
SURFACE GRAPH	X								X
SYSTAT		X	X		X				
TERRASOFT	X		X						
TURBOCAD	X				X				
VERSACAD	X				X				
VISIBLE	X			X	X				

Package name	Platform		Generic Category						
	PC	MAC	GIS	Carto	CAD	RS	Draw	Digit	Cont
ULTIMAP	X				X				
WHIZATLAS/MAP		X		X					
WHIZSURF		X							X
WINGS	X		X						

Platform:

PC	Personal computer
MAC	Apple Macintosh computer

Generic category:

GIS	Geographical information system
Carto	Cartographic system
CAD	Computer-aided design system
RS	Remote sensing/image processing system
Draw	Drawing/design package
Digit	Digitising/data capture package
Cont	Contour/surface generation package

Index

Page numbers appearing in **bold** refer to figures and page numbers appearing in *italics* refer to tables.